MEASUREMENTS FOR TERRESTRIAL VEGETATION

MEASUREMENTS FOR TERRESTRIAL VEGETATION

CHARLES D. BONHAM

Professor of Quantitative Ecology
Colorado State University
Fort Collins, Colorado

WILEY

A WILEY-INTERSCIENCE PUBLICATION

JOHN WILEY & SONS

New York • Chichester • Brisbane • Toronto • Singapore

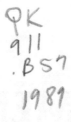
Library of Congress Cataloging in Publication Data:

Bonham, Charles D., 1937–
　Measurements for terrestrial vegetation.

　"A Wiley-Interscience publication."
　Includes bibliographies.
　　1. Plant communities—Measurement.
I. Title.

QK911.B57　1988　　581.5′247′028　　88-14269
ISBN 0-471-04880-1
Printed in the United States of America

10 9 8 7 6 5 4 3 2

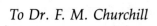

To Dr. F. M. Churchill

PREFACE

This book was developed from experiences gained over a 20-year period of vegetation measurements. Additional appreciation for the need of these measurements and insight on their quantitative characteristics were developed from courses taught by the author to both undergraduate and graduate students at Colorado State University. The book introduces the four commonly used measures of vegetation and associated units used to obtain such measures. Furthermore, a basis is provided to develop this understanding from a statistical variability viewpoint.

In general, plant community ecology has emphasized two broad categories: functional processes in plants and structural characteristics of plants. The structural approach to plant ecological studies emphasizes the form and arrangements of plants within the community and assumes that functional processes account for physical characteristics of plants. That is, the weight of individual plants, plant size and shape, leaf arrangements, and numbers of individuals all result from physiochemical processes of plants.

No attempt has been made to give more concise meanings of words and terms than those already employed for up to a half-century. These include the use of production for standing crop, double sampling for regression equations, quadrat for plot, and line transects for a measurement tape, to mention a few. The book deals with principles and procedures used to obtain structural measurements of terrestrial vegetation communities. Emphasis is placed on frequency, cover, density, and biomass as commonly defined in the preponderance of literature on such measures. A special case for measures of vegetation characteristics resides in double sampling. While any measure not obtained directly by contact with vegetation has been "double sampled," the most commonly used definition is retained in the book.

Efforts were made to balance conceptual as well as practical characteristics of measurement procedures and techniques presented. Yet, an abundance of studies are referenced whereby individual units of measures were made. A number of references are also used to illustrate differences in estimates of individual measures, such as density, obtained by various techniques and units.

I am grateful to the reviewers of the manuscript. Many suggestions greatly improved the manuscript. Finally, I am especially indebted to Dr. Dot Helm who as a graduate assistant laboriously assisted with review and abstracting of literature, to Holger Jensen who checked all literature abstracts and equations, to Tana Allshouse, who provided all typing, additional editing, and efficient management of the manuscript for completion of the work. My thanks to Doris Rust for valuable assistance on all illustrations and to the Range Science Department at Colorado State for support of the work.

CHARLES D. BONHAM

Fort Collins, Colorado

CONTENTS

1 INTRODUCTION

Variation in the morphology of plants resulted in the grouping of plants into broad categories on the basis of life-forms. Major life-forms are represented by terms such as "tree," "shrub," "grass," and "forb." These life-forms often provide a basis to describe major terrestrial plant communities (Odum 1971). Life-forms of plants or plant species can be described by a number of characteristics such as biomass, frequency, cover, and density. Some life-forms or plant species are perhaps better described by certain characteristics of measure than are others. A combination of objectives of a study and species involved will determine what characteristics are to be measured for an effective description of vegetation. To see this more clearly, consider the measurements of cover, which are estimates of relative areas that a plant controls to receive sunlight. In comparison, biomass directly indicates how much vegetation is present, and particular species indicate the amount of forage available to herbivores in the area. Density describes how many individual stems or plants occur per unit ground area, while frequency describes the dispersal or distribution of a species over the landscape. Each of these species characteristics has a distinct use in vegetation characterization and description. Often, several measures are used in combination for an in-depth description of vegetation (Bonham 1983).

A distinction is made between a unit and a technique, as used here, to discuss vegetation measures. The unit is a distinct measure of an individual component. Since frequency, cover, density, and biomass are expressed in quantities per unit of area, these are units associated with equipment. Points, plots, and tape measures are often used to obtain these units of measure. On the other hand, a technique includes the process used to obtain the unit of measure, specifically: location of the observation, clipping, observing a hit by a point, and summarization of the unit data. Thus, methods used in measurement are not units but, rather, are ways of doing things to obtain the unit of measure. Often the terms "methods" and "techniques" are used synonymously in the vegetation literature to include pieces of equipment. Thus, one reads phrases such as "the method used was a 0.5-m^2 quadrat." No distinction is made here between techniques and methods because both describe or imply a process used in obtaining a measure of given vegetation characteristics. In any case, methods or techniques involve the use of units, equipment, and field procedures to make a measurement.

Measurements of vegetation characteristics have been made for more than a century, and techniques developed to obtain these measurements are numerous. Few units and techniques are comparable, even for measuring the same characteristic of vegetation, such as cover. This is true because objectives to obtain the measure differ. Yet, comparisons of vegetation characteristics over time and space are often necessary. If comparisons are to be made, comparable methods must be used to obtain measurements.

There are basically two objectives to be considered in the selection of measurements and techniques used to obtain measurements. One objective is to assess effects of a perturbation applied to vegetation. For example, the objective may be to describe the characteristics of the native vegetation at a given time, and then again at a later date, to assess changes. Clearly, the rule of least costs will often dominate the method of evaluation and comparison of vegetation characteristics.

The second objective is to describe vegetation characteristics so that the measurement can be used as a standard or a baseline. Otherwise, all descriptions of the vegetation would be relative and not subject to valid comparisons. The two objectives are compatible and should not be considered as competing for resources, especially monetary, in order to be attained.

The purpose in a study of units and techniques to obtain vegetation measurements is twofold in nature: (1) to make proper choices of appropriate units used to estimate a characteristic of vegetation, that is, frequency, cover, density, and biomass, and (2) to properly select and utilize a sample design that will provide unbiased estimates of the characteristic measured. One will gain efficiency in carrying out both of these objectives simultaneously only through proper use of methods in the field, followed by effective data analysis procedures.

No known set of techniques is free of disadvantages for any measurement or set of measurements. Rather, selection of a technique should be made with perhaps an understanding that certain modifications may be needed to optimize its use. Modifications of techniques to reduce effects of biased estimates or to limit disadvantages are frequently described in the literature. In general, however, all techniques used to measure vegetation are related to land areas. That is, frequency, cover, density, and biomass amounts are correlated not only for certain species but also by the fact that each is estimated with reference to a land area and, in some cases, volume occupied.

1.1 HISTORICAL BRIEF

Measurements of vegetation date to antiquity. In the third century B.C., Theophrastus observed that certain relationships existed between plants and their environment. Thus, he was an early contributor to plant ecology in the quantitative sense. Still, centuries passed with qualitative assessments dominating vegetation descriptions, and plant species in particular. Geographic descriptions of vegetation occupied the interest of many naturalists in Europe, where only listings of species dominated their efforts. Indeed, lists of plants provided the beginning of vegetation characterization as a true quantitative approach. Emphasis was placed on such lists throughout the eighteenth and nineteenth centuries on the European continent. It was in Europe that Raunkiaer (ca. 1900) used the first known plot to obtain a quantitative measure of plant species: that of frequency, although the work was still to describe geography of plants (Raunkiaer 1934).

The work of F. E. Clements in the United States increased precision in vegetation measurements. Clements, in 1905, coined the word "quadrat" for use in vegetation data collection. While the term technically defines a four-sided plot, its usage over time has been adulterated to include any plot shape, even a circle.

Gleason (1920) further advanced the concept of quantitative measurements. He described applications of the quadrat method in description of vegetation characteristics. In particular, Gleason (1922, 1925) developed a thorough explanation of species and area (or space) relationships. His concepts led to the notion that sample adequacy could be determined from the relationship between the number of species encountered as a plot area increases. The well-known species–area curve is still used at present to determine relative sample adequacy.

Measurements and analysis of vegetation characteristics during the 1920s led to statistical applications to plant ecology. Kylin (1926) introduced the concept of "mean area" and defined it to be the inverse of density, that is, the number of individuals per unit of area. He was also among the first to present explanations for the relationship between density and frequency, which is of a logarithmic nature, not linear. Kylin's work followed that of Svedberg (1922) in Europe, and their approach to measures of vegetation was of a statistical nature that encouraged many others to examine vegetation characteristics from a quantitative point of view.

Cain (1934) and Hanson (1934) compared quadrat sizes, while Ashby (1935) gave an early introduction for the use of quantitative methods in vegetation descriptions and (Ashby 1936) published on the topic of statistical ecology. Bartlett (1936) gave examples of statistical methods for use in agriculture and applied biology, but Blackman (1935) had previously introduced statistical methodology to describe the distribution of grassland species. These early studies in vegetation measurements emphasized the dispersion of individuals in plant communities. Thus, patterns of dispersal were very much in the forefront of most quantitative assessments of vegetation characteristics.

How plants are arranged spatially implies distance measures and, subsequently, pattern. This emphasis on pattern analysis began in earnest in the 1920s and reached a peak in the late 1940s and early 1950s (Greig-Smith 1964). Interest in species patterns was briefly rekindled in the 1970s as distance measures were again used for density estimates (Green 1966, Beasom and Houcke 1975). Many distance measures in plant patterns use a single, linear dimension and are, thus, referred to as "plotless" methods. The return to plotless methods in the United States was essentially driven by time–cost considerations needed for large-scale inventories of forests and rangeland resources.

Since distance measurements included rigid assumptions about the distribution of individual plants, an understanding of patterns found in natural plant populations was necessary. Therefore, a great deal of effort was expended to develop acceptable modifications to plotless methods for use in the estimation of frequency, cover, density, and biomass. Thus, measurements of vegetation actually began to find a place in the work of professional plant ecologists from the 1920s onward. Very few professionals in vegetation ecology, however, mastered the seemingly more difficult merger of this discipline with that of statistics. Very few specialists study statistical plant ecology, especially with emphasis on native vegetation characteristics.

1.2 UNITS OF MEASURE

The science of measurement, which is called *metrology*, has been a vital part of science, especially the physical sciences, for centuries. The science of metrology was given much attention during the nineteenth century because a better system of units and standards for measurements is needed to assist the field of physics (Pipkin and Ritter 1983).

The metrology of vegetation itself, however, is of even more recent origin. Only within the decade 1970–1979 was there major progress on the determination of fundamental constants needed to relate measured vegetation characteristics displayed by density, cover, and so forth to biological and ecological theory.

The initiation of the International Biological Program (IBP) to study the interrelationships of organisms and their environment that operated in an ecosystem pursued the integrated systems approach. Mathematical and statistical models formulated during this time provided some fundamental insight as to how such systems of organisms functioned individually and collectively. Thus, for example, constants for energy and nutrient transfer through a system were provided, which resulted in a clearer understanding of how measured characteristics described plant–environment relationships. For example, the amount of biomass accumulation by an individual species can be used to assess that species role in nutrient utilization and recycling within the vegetation system as a whole.

Most vegetation measurements are now made in metric notation, which is used throughout this book. Table 1.1 provides a definition of the relationship that exists among linear, area, volume, and weight measures from the metric system. The volume measure is given in this form because some measures of weight of vegetation biomass may be reported as per unit volume occupied. Essential or fundamental constants used in measure-

TABLE 1.1   Metric Weights and Measures

Linear Measure

10 millimeters (mm)	= 1 centimeter (cm)
10 centimeters	= 1 decimeter (dm)
	= 100 millimeters
10 decimeters	= 1 meter (m)
	= 1000 millimeters
10 meters	= 1 dekameter (dam)
10 dekameters	= 1 hectometer (hm)
	= 100 meters
10 hectometers	= 1 kilometer (km)
	= 1000 meters

Area Measure

100 square millimeters (mm^2)	= 1 square centimeter (cm^2)
10,000 square centimeters	= 1 square meter (m^2)
100 square meters	= 1 are (a)
100 ares	= 1 hectare (ha)
	= 10,000 (m^2)
100 hectares	= 1 square kilometer (km^2)
	= 1,000,000 (m^2)

Volume Measure

1000 cubic millimeters (mm^3)	= 1 cubic centimeter (cm^3)
1000 cubic centimeters	= 1 cubic decimeter (dm^3)
	= 1,000,000 (mm^3)
1000 cubic decimeters	= 1 cubic meter (m^3)
1,000,000 cubic centimeters	= 1,000,000 (mm^3)

Weight

10 milligrams (mg)	= 1 centigram (cg)
10 decigrams	= 1 gram (g)
	= 1000 (mg)
10 dekagrams	= hectogram (hg)
	= 100 (g)
10 hectograms	= kilogram (kg)
	= 1000 (g)
1000 kilograms	= 1 metric ton (t)

ments of vegetation may include conversion from one system to another. For example, in order to interchange units from the metric to the English system, constants are needed. Approximation values for some of these constants are given in appendix Table A.1 (appendix at end of the book).

Additional constants to those in Table 1.1 are often needed in vegetation measurement work. Such constants are given in appendix Table A.2 since these small areas are often used to estimate weight of plant biomass, especially that of forage, in much of the world. Some public land agencies in the United States still use "lb/acre" for management of forage resources, yet use metric dimensions for obtaining these estimates. Therefore, conversions are given for the most commonly used plot areas.

Calculation of a constant for any given area to a larger area is as follows:

$$\text{unit of wt(1)/large area} = \frac{\text{one unit of wt(2)/large area}}{\text{(units of wt(2)/unit wt(1) (small area)}} \quad (1.1)$$

That is, for a 0.25-m² plot (appendix Table A.2),

$$\text{kg/ha} = \frac{\text{g/10,000 m}^2}{(1{,}000 \text{ g/kg})(0.25 \text{ m}^2)} = \frac{10}{0.25} = 40$$

Still other constants may be useful for conversion of one or more measures into another measure. For example, cover percentage of a species may be used to estimate biomass weight (grams) of the species, in which case the constant is estimated by least squares procedures of regression analysis. The general form of the equation is usually

$$\text{Biomass (g/area)} = f\left[\sum_{i=1}^{k} (\text{measure } i)\right] \quad (1.2)$$

where i is one or more independent measures such as cover, stem diameter, and so on, and k is the number of independent measures made on the plant. The function f also includes measurements in the equation as given in appropriate chapters of the book.

1.3   CHOICE OF TECHNIQUE

Selection of an appropriate technique to use for obtaining a measure is based on several criteria. Emphasis in this chapter is based on two major characteristics: (1) those involving the physical aspects of vegetation and (2) those involving biometrics and econometrics of the techniques.

1.3.1 Vegetation Characteristics

Selection of proper units and measurement techniques to use requires a knowledge of floristic composition of the vegetation type. That is, units to estimate density, biomass, and cover are determined by the plant community life-forms. The abundance of these measures determines the size of quadrat, the specific distance measure, and the number of sampling points needed. In general, dense vegetation, which usually implies higher density of individual plants, larger plants, or both, can be measured with fewer large plots or fewer points than a sparsely occupied area. The latter areas often require more points, plots, and so on because the variation is larger for the measure, such as cover or biomass.

Life-forms present in a vegetation type are suggested as a consideration in selection of technique. But sometimes a given form, such as the shrub or tree form, will have more variation in size over species than will grasses or forbs in a vegetation type. The same can be true for forbs compared to grasses. That is, size variation among species within a life-form can suggest the use of a technique such as the size of a quadrat for individual plant counts or biomass determination. If a large plant is abundant and may fully occupy the quadrat when encountered, then spatial exclusion occurs for all other species in that case; in other words, no other individual can occur in a quadrat occupied by another large individual. In such cases, underestimates of a measure for some species may result, while other species, especially large ones, may be overestimated.

Patterns of distribution for individuals within a species, and patterns for species, are of major concern in technique selection and use. If all distributions were random, or very nearly so, then measurements would not yield biased results in most cases. Random distributions imply that no pattern is present in the measure obtained and that such measures can be obtained from a random sampling process to provide unbiased estimates of the measure. In vegetation measurements, the presence of patterns in the measure causes the greatest biometric concern, which, in turn, influences the economics of measurements.

1.3.2 Biometrics and Econometrics

Biometrics (the science of statistics applied to biological observations) and econometrics (the science of statistics applied to economic data) are useful in the selection of appropriate measurement techniques. Statistics provide estimates of the population mean and an estimate of its variance with assignable probabilities for confidence limits. When the data distribution form (e.g., normal or binomial) is known or assumed for a measure, a technique that provides the smallest value of an estimate of the variance of the mean is the best estimate of the mean. Furthermore, such a technique provides the minimum number of observations to be taken for that given measure of biomass, cover, density.

Obviously, the foregoing discussion implies that certain techniques will also be more efficient from the economic viewpoint. A technique that requires more observations than another to obtain the same precision is not as cost-effective if the cost is the same for each observation. Additionally, accuracy of the estimates is to be considered. In some cases, a technique may be precise, that is, gives repeatable results. However, it may not be accurate, which means that the technique does not estimate the true population value very well.

Emphasis should be placed on the sample error of a technique. This error is not a mistake or an oversight, but rather involves the variation present in actual measures (cover, density, and biomass). An estimate of this error is usually defined by the "standard error" of the mean. The magnitude of the sampling error depends on the (1) number of observations, (2) inherent variability of the measure, and (3) method of selecting a sample. That is, location of units in the field is made according to random or other methods. All of these aspects of the sample error will affect the cost of sampling.

1.4 VARIATION IN VEGETATION

Sources of variation in measures of vegetation characteristics are many. Vegetation characteristics of frequency, cover, density, and biomass are affected by species life-form; species composition; seasonality; previous use by humans and animals; and edaphic and climatic characteristics. Already, life-forms have been suggested as a variable that affects the selection of a technique used for a measure. Life-form (tree, shrub, herb) is a source of variation in vegetation in terms of its relationship to frequency, amounts of cover, number of individual plants possible in an area, and effects on biomass of a plant.

Precipitation, air and soil temperature, soil moisture, and time in relation to initiation and cessation of plant growth all affect the measure of a characteristic for a certain species. More importantly, however, they contribute significantly to variation of the measure when made over all species present. For example, total biomass of an area depends on the phenological stages of the species present. Variations among time intervals over a growing season or years, then, will depend on the development stage of both major and minor species. Previous use of, or destruction of, vegetation is often reflected in the variation found in vegetation characteristics. Previous harvesting of trees, heavy grazing by large herbivores, periodic infestation of insects, and disease are significant sources of variation in measures of vegetation. Variations are often noted also by species composition, which indicates a secondary or greater successional stage of the plant community. Edaphic (soil) sources, as contributors to variation in vegetation, include parent material, stage of soil development, and physical and chemical properties of the present soil.

While weather is a measure of present events such as air temperature, wind, and precipitation, climate is a long-term phenomenon. Weather, in general, affects the measure of vegetation cover more so than it does density for perennial species. However, annual and ephemeral species are affected through expressions of frequency, density, cover, and biomass by weather events occurring within a few days. Climate, on the other hand, introduces variation in vegetation through its determination of species composition and reproduction in perennial species.

These sources of variation should be used to stratify for sampling purposes to enhance efficiency in the sampling process. A "stratum" is a unit of vegetation that is homogeneous with respect to species composition, soil type, and so on. Or a combination of vegetation and environmental characteristics are used to form strata. Strata should be used to classify measurements to ensure minimum variability within a stratum and maximum variability among strata for measures of interest. Vegetation typing accomplishes stratification on a large scale, but much more efficiency in sampling is gained within a stratum found in the type that is more homogeneous. Then, statistical procedures provide a more efficient estimate, both statistical and economical, of the measure. Measures of vegetation will have less variation if like life-forms are placed into strata according to trees, shrubs, and herbaceous plants. Season of growth for herbaceous plants should be stratified into early, mid, and late season, while kinds of past use and intensity of use also provide strata. Table 1.2 provides a schematic diagram for relationships among vegetation variability and general effects on measurement attributes.

TABLE 1.2   Relations of Vegetation Characteristic Variability and Measurements of Cover, Density, and Biomass

To Large	Variability	From Small
←	Sample size increases	
←	Plot size increases	
←	Need for stratification increases	
←	Effectiveness of double sampling increases	

1.5 OBSERVATIONAL UNITS

A unit is a distinct, discrete member of a population that can be analyzed in an aggregate or as a whole. It is any quantity (weight, percentage, length) used as a measurement and is representative of the population as a whole. Then, amount of plant biomass in an area of a plot is a single unit, just as a point touching a leaf is a unit of cover. Thus, observational units are the various pieces of equipment used to obtain measures of vegetation characteristics. Some literature references prefer to use the term "sampling units," but units on which observations are made appear to be more appropriately called "observational units." The equipment used as observational units is described in Chapter 2.

1.6 SAMPLING

Sampling is best understood when compared to a complete enumeration or census. Both of the latter refer to measurements of every individual unit in a population. For example, every possible quadrat or distance between plants constitutes a population. It is well known that a sample from a population of measurable units will essentially contain the same information as a complete enumeration, yet will cost far less than the census. What is less well known, though, is the fact that sample-based data may be more reliable than a 100% inventory. This follows from the fact that samples are often taken with greater care than can be used in a complete census because more expertise can be used in sampling. Sample-based data also can be collected and processed in a small fraction of the time required for a complete census.

The objective for sampling is to obtain an unbiased estimate of the population parameters: the mean and its variance. Samples are made up of a number of observations. Each sample is of size n, the number of observations made for the measure of interest. Then, each sample has a set of observations specific to it. Therefore, the word "sample," in the statistical sense, refers to a set of observations, not a single observation. An observation, in contrast, is a single measurement made from a "unit," which was defined in Section 1.5. Usage of these terms should follow these definitions to prevent misinterpretation of sample size information.

An understanding of basic statistical concepts should be developed before fieldwork is begun, regardless of the reason for obtaining measurements. For this purpose, principles of statistics are given in elementary detail in Chapter 3. Of particular importance is the relationship of variation and sample size needed to adequately describe population measures.

Concepts of accuracy and precision are important in the selection of a measurement unit. For example, some units and techniques are precise (results are repeatable) but not necessarily accurate (providing true population value). The difference between the two emphasizes the sample estimate (precision), while "accuracy" refers to the true population measure. A sample-based estimate from a unit and associated technique may be both precise and accurate. An example is a large number of randomly located points to estimate cover.

Units used to measure characteristics of vegetation influence the data distribution. For example, the size of a plot determines the number of individuals that can occur within the plot boundaries. Then, the Poisson distribution can be derived from consideration of plot area relative to the total area sampled. No derivations are given, only explanations of the relationships of measurements to distributions and relationships among distributions. The Bernoulli distribution is discussed to develop an intuitive basis for occurrence of the binomial distribution. The point-and-count binomial distributions are given with parameters and associated sample size adequacy as estimated from an equation based on the normal distribution.

Data distribution form determines available transformations that are appropriate to use. Specific transformations for data include the logarithmic, square-root, and angular. These follow from the fact that data display certain characteristics and form a frequency distribution with known parameters.

1.7 FREQUENCY AND COVER

Frequency is the percentage of a species present in a sample unit and, therefore, it is influenced by size and shape of sample units. Frequency can be estimated by plots, points, and lines. Frequency is a useful index for monitoring changes in vegetation over time and comparing different plant communities. Mathematical relationships may be used to estimate plant density and cover from frequency data. The index is also highly sensitive to abundance and pattern of plant growth. Selection of an appropriate plot size and shape requires preliminary study of the vegetation type. Some plant ecologists recommend plots of certain size on the basis of experience, while others calculate plot sizes from mathematical relationships between some vegetation characteristics for statistical considerations. The scope of frequency and details for calculation of plot size and other considerations are discussed in Chapter 4.

Cover can be measured for vegetation in contact with the ground (basal area) or by projected aerial parts of vegetation onto the ground (foliage cover). All methods of measuring cover depend on interception of the plant by a quadrat, the area of which may be very close to zero (0) for a point and yet is still two-dimensional. In fact, all techniques give cover estimates on a two- or three-dimensional basis. Yet, a line is often assumed to be one-dimensional when, in fact, its width is very narrow while its length is very long.

Points have been used singly, as in the step-point method, or in frames, as in the point-frame method. Spacing of pins within a frame is, in general, closer for intensive studies, and farther apart for general surveys. This is so because individual species are of more interest in the former, while a general cover estimate will suffice in the latter. Cross-hairs in telescopic sights are used for sighting the points rather than the lowering of a pin. The line-point method involves point readings along a transect and may be faster than the line intercept but does not work as well when cover is small in value—less than, say, 15%.

Line-intercept methods involve laying out a transect and measuring the length of a species intersected. Objectives of the study determine whether to measure basal area or foliage cover for each species. Line-intercept methods give values closer to true cover as calculated from ellipse formulas than with variable plot and loop methods. Line-intercept data are more accurate, and data are obtained more rapidly than by the use of quadrats in communities with different-sized individuals of plant species.

Quadrats of varying sizes have also been used to measure cover. Earliest methods included charting areas covered by plants and density lists, neither of which are used anymore. The most often used quadrat methods involve ocular estimates of percentage cover by species. This is usually accomplished by cover classes and use of the midpoint value of the cover class for data analysis. Use of cover classes enables repeatable estimates to be made by different observers, but the data cannot be analyzed by standard statistical methods.

Plotless techniques have been used to estimate cover, but not extensively—specifically, because the methods are difficult to use when cover is greater than 35%. Plotless techniques include distance measures and other unbounded units. But plotless techniques are more efficient than other methods (bounded as a quadrat) in open shrub and forested communities.

Shrub cover and tree cover have been determined with the line-intercept method in dense stands and in open stands (<35% cover) by variable-plot (plotless) methods. Canopy coverage of species life-forms is also measured to indicate the amount of sunlight interception. Chapter 4, on frequency and cover, gives examples and comparisons of units used to measure plant cover.

1.8   DENSITY

Density measurements require that individual plants be countable. Density, historically, has been an important measurement for trees and shrubs and somewhat less important for bunchgrasses and forbs. Essentially, the counting process is slow and tedius for herbs and is questionable for use with sod-forming grasses and other growth-forms where the identity of an individual plant may be difficult to establish. However, density can be estimated if a standard individual is defined properly.

Density should be used for characterization of herbaceous species in spite of drawbacks because the measure remains more constant from year to year than will measures of cover or biomass. Herbaceous densities are usually obtained by counting individuals in a quadrat, while variable-plot methods are often used for woody species. Square quadrats have been suggested for homogeneous stands and a line strip quadrat for heterogeneous stands. Plotless techniques to estimate density include point-centered quarter (PCQ), wandering quarter, line transect, random pairs, nearest neighbor, and closest individual. Chapter 5 provides details of methods and comparisons of methods used to estimate plant density.

1.9 BIOMASS

Biomass is another primary vegetation measure because it indicates the quantity of resources, such as water, used by a species in the community. Biomass is also a measure of how much of the community's resources are tied up in different species. The primary method used to measure herbaceous biomass is to clip or guess amounts of biomass in quadrats. Tree and shrub biomass is usually estimated from dimensional analysis. That is, other measures, such as plant height and crown diameter, are made and used to predict biomass from regression equations.

In some studies of quadrat size and shape, certain sizes of rectangles and circular plots are the most expedient for clipping. Other sizes of circular and rectangular quadrats are statistically more efficient. Historically, long narrow plots, placed across the direction of change, were generally considered the most efficient, but recent studies contradict this view. Variation does not simply follow a single direction but is two- or three-dimensional in the field.

Measurements of cover and frequency of plants have been used successfully to estimate biomass production of some species with correlation coefficients exceeding 90%. Generally, efforts such as the latter fail to produce an unbiased estimate, simply because proper statistical procedures are not followed. Usually the lack of randomization in the sample process will lead to biased estimates.

Optimum plot sizes differ for determination of biomass of forbs, grasses, dead material, and total green material. Additionally, optimum size and shape of plot will differ among vegetation types. Furthermore, sample adequacy, in terms of observations needed, also varies according to the size and shape of the quadrat, species combination, and vegetation type. To appreciate why this is true, recall the sources of variation in vegetation characteristics presented in Section 1.4.

Biomass of shrubs is usually measured with two major components in mind: total biomass and forage for animals. If the objective of a project is an ecological characterization of the vegetation, then only total biomass need be considered. Otherwise, if large herbivores are present and feed on shrubs, then forage (or "browse" as it is often termed) is measured. Shrub and tree biomass and/or current annual growth are determined by indirect methods using dimensional analysis. As previously suggested, the most frequently used dimensions include crown diameter, crown area, height, and basal stem diameter. Many of these dimensions involve data transformations before acceptable estimates are obtained.

Shrub foliage estimates and browse are based on twig length and count measurements. Then, regression equations are computed by size classes for shrubs and vegetation types. In fact, very few studies have shown meaningful relationships between tree and shrub size and biomass unless plant size classes by site are considered. Site differences probably account for growth differences and express environmental differences directly. Suggested measurements for double sampling of shrubs and trees are crown length and width, plant height, basal stem diameter, and percentage of live crown cover. The percentage cover may not be useful for indirect estimation of total biomass, but in species with large patches of dead crown, cover provides an acceptable estimate of biomass when used in regression equations. Chapter 6, on biomass, provides details.

1.10 MONITORING AND EVALUATION

Objectives of vegetation measurements differ with respect to various project goals. One may be interested in mapping and the description of vegetation, studying the patterns of primary or secondary succession, or interactions of soil, plant, and water relationships. The objectives and methods of vegetation measurement may be different, but the plant characteristics to be measured remain the same: frequency, cover, density, and biomass. Once information is available on these characteristics, the data can be synthesized and manipulated in a manner that best suits the objectives of the study.

An essential prerequisite of any vegetation measurement for inventory purposes is general familiarity with the area and recognition of general physiognomy of the vegetation. This can be achieved through a reconnaissance survey of the area. Topographic maps, soil survey maps, and aerial photographs are very helpful in such a survey. The topography, aspect, and soil influence vegetation composition and plant characteristics. It is, therefore, possible to map ecological units or vegetation types on survey maps. Once the ecological units are delineated, the next step is to choose a sampling design and measure plant characteristics of interest. A summary of the sampling statistics should be used in a narrative description of the ecological sites.

There are few areas not grazed by domestic or wild animals. In fact, competition between domestic and wild animals is significant in many parts of the world. Moreover, herbivorous insect populations sometimes reach epidemic proportions, destroy native vegetation, and, thereby, reduce forage resources for animals. Even at normal population levels, small herbivores, although sometimes not visible, consume substantial quantities of forage that would otherwise be available for large herbivores (Caperina 1987).

Methods of determining carrying capacity for herbivores range from general reconnaisance to detailed surveys. The choice depends on objectives and resources available to undertake such a survey. Simple information on herbage biomass is not sufficient basis for a decision regarding the number and kinds of herbivores that can graze on a given ecological site or vegetation type. Such a decision depends on the successional status of vegetation and the condition of the vegetation. Vegetation is considered to be in excellent condition when a vegetation type represents a climax stage in plant succession or when the most productive stage is the subclimax stage. By comparison to different seral conditions, vegetation condition is designated as excellent, good, fair, or poor. The objective is usually to maintain vegetation resources in at least a good condition for grazing of large herbivores. Monitoring the change of vegetation condition enables adjustment of land use practices. Planning and management decisions for vegetation use should be based on accurate assessments of the vegetation resources, and such assessments should be regularly updated to detect trends and revise the management plans accordingly. Methodology for vegetation monitoring and evaluation is given in Chapter 7.

1.11 OVERVIEW AND SUMMARY

Processes involved in measurements of vegetation characteristics and associated uses with environmental data are given in Figure 1.1. Variability encountered in measurements should not be determined solely after data have been collected. Even casual observations, made visually, reveal sources of variability likely to influence the measures made. The most obvious sources of variation are usually those influences on vegetation caused by topographic, edaphic, and elevational differences among vegetation types. These same influences are also found in a micro manner within a vegetation type. Thus, one recognizes micro-topographic influences, and micro-relief differences, as well as associated soil differences. Stratification of these sources of variation into homogeneous units, beforehand, will be more efficient statistically and economically.

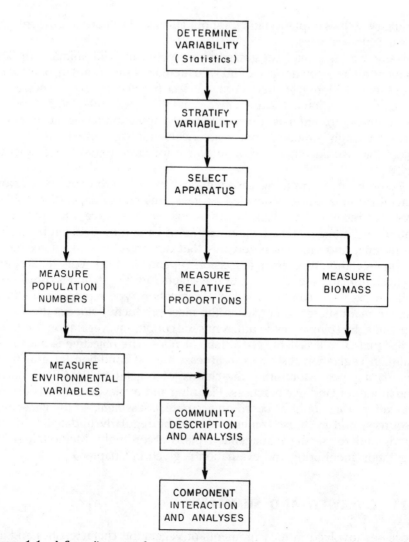

Figure 1.1   A flow diagram of processes involved in vegetation measurement and data analysis.

Selection of an apparatus to use for measurement will be influenced by the manner in which stratification of the vegetation was accomplished. For example, topography characteristics may exclude the use of certain angle gauges since an assumption is that accuracy of estimated values depends on presence of level terrain. Inclusion of strata characteristics as a consideration in selecting an observational unit is assumed in particular chapters on cover, density, and biomass. No known studies have addressed this issue, so the reader is encouraged to include the effects of vegetation variability caused by associated environmental characteristics when a selection of equipment and techniques is made.

All three characteristics of vegetation structure, cover, density, and biomass of species should be used to describe the vegetation of an area. As implied in Figure 1.1, proportions include both cover and frequency. These measures are then used to develop a complete community description when environmental data are also available. Basic soil descriptions, along with general meteorological data for the area, are sufficient to provide an analysis of general interactions occurring in vegetation of an area. That is, a vegetation–environmental systems analysis can be used to develop an interpretation for general uses or perturbations that might occur to the vegetation.

1.12 BIBLIOGRAPHY

Ashby, E. 1935. The quantitative analysis of vegetation. *Ann. Bot.* **49:** 779–802.

———. 1936. Statistical ecology. *Bot. Rev.* **2:** 221–235.

Bartlett, M. S. 1936. Some examples of statistical methods of research in agriculture and applied biology. *Suppl. J. Roy. Stat. Soc.* **4:** 137–183.

Beasom, S. L., and H. H. Haucke. 1975. A comparison of four distance sampling techniques in South Texas live oak mottes. *J. Range Manage.* **28:** 142–144.

Blackman, G. E. 1935. A study by statistical methods of the distribution of species in grassland associations. *Ann. Bot.* **49:** 749–777.

Bonham, C. D. 1983. Field methods for plant resources inventories. In *Resource Inventory and Baseline Study Methods for Developing Countries,* F. Conant, P. Rogers, M. Baumgardner, C. McKell, R. Dasmann, and P. Reining (eds.). American Association for the Advancement of Science, Washington, DC, 539 pp.

Cain, S. A. 1934. Studies on virgin hardwood forest. II. A comparison of quadrat sizes in a quantitative phytosociological study of Nash's Woods, Posey County, Indiana. *Am. Midl. Nat.* **15:** 529–566.

Caperina, John L. (Ed.). 1987. *Integrated Pest Management on Rangeland—a Shortgrass Perspective.* Westview Press, Boulder and London, 426 pp.

Gleason, H. A. 1920. Some applications of the quadrat method. *Bull. Torrey Bot. Cl.* **47:** 21–33.

———. 1922. On the relations between species and area. *Ecology* **3:** 158–162.

———. 1925. Species and area. *Ecology* **6:** 66–74.

Green, R. H. 1966. Measurement of non-randomness in spatial distributions. *Research, Populations, Ecology* (Kyoto) **8**(1): 1–7.

Greig-Smith, P. 1964. *Quantitative Plant Ecology,* 2nd ed. Butterworths, London, 256 pp.

Hanson, H. C. 1934. A comparison of methods of botannical analysis of the native prairie in western North Dakota. *J. Agric. Res.* **49:** 815–842.

Kylin, H. 1926. Über Begriffsbildung and Statistik in der Pflanzen-Soziologie. *Bot. Notiser.* (1926) 81–180.

National Research Council/National Academy of Sciences. 1984. *Developing Strategies for Rangeland Management.* Westview Press, Boulder, CO. 2022 pp.

Odum, E. P. 1971. *Fundamentals of Ecology,* 3d ed. Saunders, Philadelphia, 574 pp.

Pipkin, F. M., and R. C. Ritter. 1983. Precision measurements and fundamental constants. *Science* **219:** 913–921.

Raunkiaer, C. 1912. Measuring apparatus for statistical investigations of plant formations. *Bot Tidsskr.* **33:** 45–48.

Raunkiaer, C. 1934. *The Life-Forms of Plants and Statistical Plant Geography.* The collected papers of C. Raunkiaer, translated into English by H. G. Carter, A. G. Fansley, and Miss Fausboll, Clarendon, Oxford, 632 pp.

Svedberg, T. 1922. Ett bidrag till de statistika metodermas anvandning inom vaxtbiologien. *Svensk. Bot. Tidsskr.* **16:** 1–8.

2 UNITS FOR MEASUREMENTS

There are many units (pieces of equipment) available to obtain measures of vegetation characteristics. Furthermore, numerous modifications have been made to basic units to obtain more efficient estimates of a given measure. Historically, most units were developed and used for a specific purpose and statistical considerations were not included. As statistical procedures became important in ecological studies, unit modifications proliferated. Not all such modifications suited the criteria of providing unbiased estimates of vegetation measures. Yet, some of these latter units are among the most popular in use by vegetation ecologists. Only techniques and equipment used to obtain a unit of measure are discussed in this chapter. Specific uses, with examples, are given in chapters on measures of cover, frequency, density, and biomass.

The definition and explanation of a random variable are presented in Chapter 3. It is precisely this idea that needs to be understood before one can make a good choice of an observational unit to use in order to obtain a given measure. A random variable acquires meaning as the weight of plant (grams) per unit of area, percentage cover of a species, or numbers of individual plants per unit of area. Then one should understand the characteristics of the variable before a unit is selected to make a measurement. For example, the number of individual plants that can occur within an observational unit of a given size is limited by the size and spacing of the plants. Therefore, the size of a plot must be determined with these two characteristics of plant species in mind.

Variability of the measure is also an important consideration when a measurement unit is selected. For instance, a rectangular-shaped unit will cause more variation to occur between units rather than within a unit. Of course, this is true only if the unit is large enough to include a large range in values of the measure. This will occur if the unit's long axis is placed parallel to the field direction of variation for the measure.

All units used to measure vegetation characteristics have both advantages and disadvantages. No one unit will provide the greatest efficiency in terms of statistics, economics, and ease of use by the observer. Relative efficiency comparisons among units are often made by statistical procedures. For example, variances of the measure are estimated from several observations from each unit (quadrat, point, etc.). The unit that provides the least variance of the measure is the best one from a statistical point of view. This follows from the fact that all estimates of the measure are assumed to be unbiased estimates of the mean. Then, the unit of measure with the lowest variance of that mean also gives the shortest confidence interval for the estimated mean. Thus, precision is greater when the unit providing the least variance is used to obtain, for example, an estimate of cover.

2.1 COVER

The oldest method used to measure vegetation cover is the point. The contact of a point on a vegetation part has been used extensively to measure cover. Individual points are used, as well as several points collected into clusters. Individual points are represented by tips of sharpened pencils, a length of stiff wire sharpened to a point, and cross-hairs in telescopes or reflecting mirrors. More recently, areas bounded by plots have been used extensively to obtain estimates of cover by species, total vegetation, litter, rocks, and bare ground. All kinds of modification in the uses of points and plot areas have been made to obtain more efficient estimates, statistically and economically. Only major units and respective modifications are presented in this chapter. Examples are given in Chapter 4 for units used to measure cover.

2.1.1 Points

Points are considered to be the most objective way to estimate cover. Points are plots with a very small area. There is minimal error for personal bias when a point is used. Either the point contacts a part of the plant, or it does not. The point, or a collection of points, is universally used for cover. Points provide for rapid data acquisition and, thus, are often economical to use. However, errors can occur from other sources such as movement of plants by wind or improper lowering of the pins by the observer.

A single point, as previously mentioned, can be defined as the intersection of cross-hairs (Fig. 2.1) or as the end of a sharpened steel pin. Cross-hairs have been used to obtain vertical hits under tree and tall shrub canopies. The instrument shown in Figure 2.1 is complicated to construct but was used before sighting telescopes were adopted. The sighting telescopes have not been used extensively to obtain plant cover because they require a stabilizing holder, such as a tripod. A straight line of sight is difficult to obtain for overhead readings in overstory vegetation. These instruments are also more expensive to purchase than other types of point. On the other hand, the apparatus illustrated in Figure 2.1 is awkward to use because at least two spirit levels are required to ensure that a 90° angle is maintained with the ground for each sample point projected vertically onto the vegetation; either above or below.

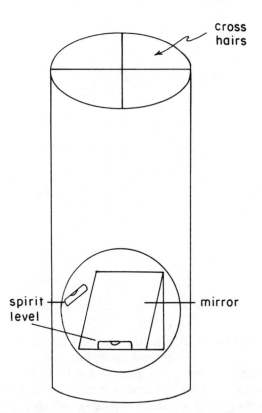

Figure 2.1 Sighting tube with cross-hairs as a point (From "A study of two communities in New Jersey ... " by M. F. Buell and J. E. Cantlon, *Ecology*, 1950, *31*(4), 569. Copyright © 1950 by The Ecological Society of America. Reprinted by permission.).

2.1.2   Point Frames

Individual points are seldom used to obtain vegetation cover estimates. Rather, points are collected into clusters or groups. One method of grouping is called a point frame (Fig. 2.2). Usually 10 pins make up the group, but any number can be used. If 10 pins are used, percentage cover can be estimated within 10% intervals since each pin is one-tenth of the total hits possible. Twenty pins provide for estimates to be within 5% intervals. Pin spacing is arbitrary and can be made to suit the growth pattern of the species. However, pins should not be so close in spacing that all 10 pins record hits on sod-forming grasses or all points make ground hits between the clumps of the grasses. Vegetation, and in particular species, patterns determine cover results which are discussed in Chapter 4.

Figure 2.3 illustrates a hit where only the first hit is recorded. Pin *a* has made contact on a plant leaf first, and additional hits would be recorded

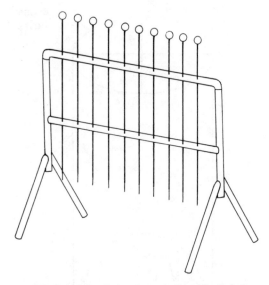

Figure 2.2   Point frame with 10 pins.

if the pin were to be pushed further toward ground level. The first hit encountered by pin *b* is the ground itself. First hits are often recorded so that total vegetation coverage of ground can be estimated. Multiple hits for a given pin are used for leaf area of species and sometimes for species composition estimates. If an individual pin is selected as a sampling process, then only one pin in the frame is used at an observation point.

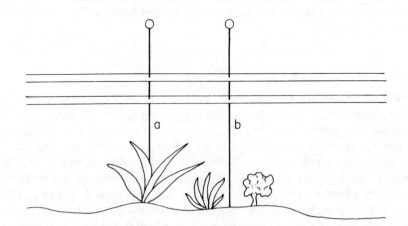

a = first hit on plant

b = first hit is at ground level

Figure 2.3   Pin *a*, first hit on plant; pin *b*, first hit on ground.

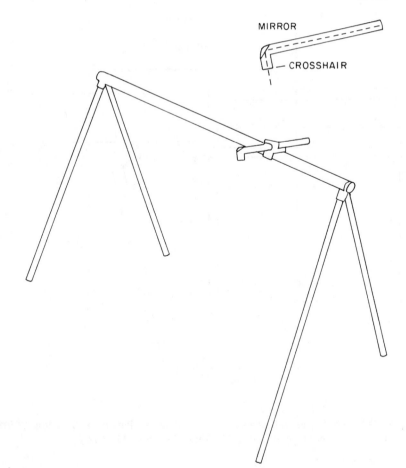

Figure 2.4 Telescope with mirror and cross-hairs for 90° sighting (From Morrison and Yarranton 1970. Canadian J. Botany 48: 296).

A telescopic sight with cross-hairs can also be mounted such that a frame of points can be obtained. Mounting of the scope onto a frame, such that the scope is movable along the bar, provides opportunity to obtain a variable number of points in a cluster (Fig. 2.4). Note the use of a reflecting mirror attached to the scope to obtain the angle needed. Angles vary with species composition; broadleaf plant cover requires less than a 90° angle for estimates, while narrowleaf plant cover estimates are more accurate with an angle of 90°. Chapter 4 gives examples and comparisons of data obtained from points with various angles used.

An illustration of an adjustable point-frame to obtain any angle between 0 and 90° is provided in Figure 2.5. Individual pins can be inclined by either using a single pin in the frame or by construction of a frame to hold only one pin. Several materials have been used to hold a pin in place. Rubber

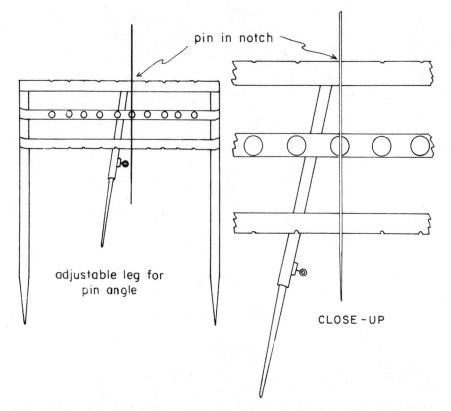

Figure 2.5   Point frame with adjustable pin inclination. Pins held by magnets (From Neal et al. "A magnetic point frame." J. Range Manage. 22: 202.).

bands can be placed around the pin and the frame to create tension so that the pin remains in the raised position until needed. Styrofoam also can be used, but it wears down quickly. A magnet, placed between each pin and the frame, may be the best way to hold the pin at any given height.

2.1.3 Point-Line

A pin placed at any distance along a tape is also considered as point sampling. However, a pin is not always necessary because the mark of distance on a tape can be used as a point (Fig. 2.6c). If the tape is placed at the vegetation height, then the latter point is easily read as a hit or a no-hit. Otherwise, a pin of some length needs to be lowered from the tape mark until contact is made either with vegetation or the ground. The pin should be lowered at a constant angle, or bias will occur. In this case, a frame to steady the pin should be used at each line point. Spacing between points along the line affects results and is discussed in Chapter 4.

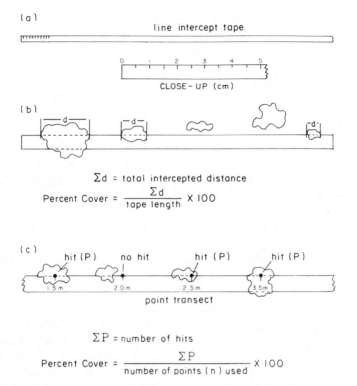

Σd = total intercepted distance

$$\text{Percent Cover} = \frac{\Sigma d}{\text{tape length}} \times 100$$

ΣP = number of hits

$$\text{Percent Cover} = \frac{\Sigma P}{\text{number of points (n) used}} \times 100$$

Figure 2.6   Line-intercept tape (*a*), intercept on plants (*b*), and point transect (*c*).

2.1.4 Line Intercept

Vegetation cover estimates can be obtained from a line placed to contact plants (Fig. 2.6*a,b*). The line or tape is stretched taut at a height to contact vegetation canopy. If basal cover area is needed, then the line is placed at ground level. The length of each intercepted plant part is measured (Fig. 2.6*b*). The length of line and total length intercepted by vegetation are used to estimate percentage cover. This method is also used to estimate species composition, that is, the percentage of total intercept made by individual species.

Tapes are often used for lines, and only one edge is read for intercept. Lengths vary, but 6- and 30-m lengths have been used regularly to study vegetation on public lands in the United States. For general surveys, other lengths of 10, 20, 50, and 100 m are used to measure cover by both intercept and line point. Locations of the lines are permanently marked by steel stakes so that monitoring can be conducted over a long period of time. Some studies have used lengths of tapes up to several hundred meters. Lengths used depend on the type of vegetation. In general, cover in her- baceous communities can be estimated with short lines (less than 50 m), while long lines (50 m or greater) should be used in some shrub and tree communities. Lines do not have to follow a 180° angle. A few studies have shown that a 90° angle can be used to obtain cover estimates in certain forests. The line length is divided into half, and the second half continues at a 90° angle to the first half.

2.1.5 Areas

Estimates of cover made from units with given areas are often biased. Bounded areas of ground surface are formed by placing a quadrat of given area on the surface. Sizes of plots used range from a fraction of a square centimeter to several square meters. A popular plot used by range ecologists in the United States is the loop, which can be either 1 or 2 cm in diameter (Fig. 2.7). These loops are placed along intervals of a tape and checked for the plant that occupies the most area. That plant, only, is recorded and all such observations are tallied for a given number of placements. Totals are used to estimate relative cover. Bias can be reduced if estimated percentage area of the loop occupied by the plant is recorded.

Cover estimates are obtained from larger areas than the loops just mentioned, as shown in Figure 2.8. This plot size and shape is the most widely used among vegetation ecologists working in herbaceous communities. The division into percentage intervals is accomplished by painting lengths on the frame. Imaginary internal boundaries are formed by the observer.

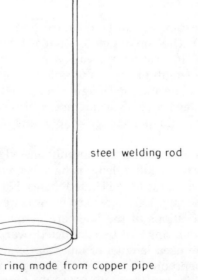

steel welding rod

ring made from copper pipe

Figure 2.7 Loop plot with diameter of 1–2 cm.

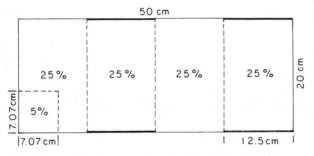

Figure 2.8   Plot used for visual cover estimates.

Further division of the quadrat areas is obtained by cutting a piece of cardboard an area equal to 1% of the total area. The cardboard piece is placed in the plot at each observation point as a guide for the observer. Then, estimates of cover are made in 1% intervals. A larger square or rectangular plot can be subdivided into smaller areas (Fig. 2.9). Usually a subunit area is 1% of the large plot area. Thus, estimates are made more objectively.

2.1.6   Plotless Units

Units without fixed boundaries that are used to obtain a measure of vegetation cover are called "plotless." This designation is used because the standard method to measure cover depends on an area defined by a fixed boundary. Plotless techniques employ units of measure which have imaginary and variable boundaries of some fashion. Probably, the best known piece of equipment used in a plotless way is the angle gauge (Fig. 2.10a).

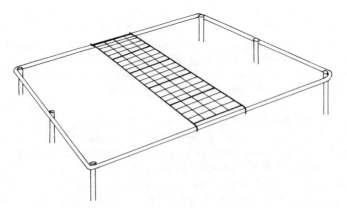

Figure 2.9   Meter square plot subdivided into small subunits that are 1% of plot area.

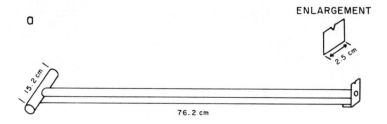

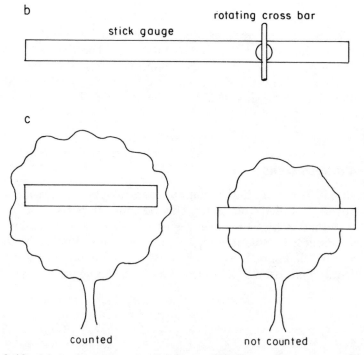

Figure 2.10 (a) Angle gauge plot (b) Slope corrections gauge (From Meeuwig, Reprinted from the Southern J. of Applied Forestry, Vol. 1, No. 3, August 1977) (c) Use of the gauge (From Cooper. 1957. "The variable plot method for estimating shrub density." J. Range Manage. 10: 113.).

The concept and geometric basis for the angle gauge were developed for use in forest inventory by W. Bitterlich of Germany in 1948. The gauge is variously referred to, but "variable-radius" sampling is a common reference, along with "Bitterlich stick." However, there are so many variations in the dimensions of length and the associated crossbar that the latter designation should be dropped unless the original linear dimensions are used.

The angle gauge is based on the fact that a small circle (a plant canopy) occupies 1% of the area of a larger circle if the radius of the large circle is k times the diameter of the small circle. If

$$k = \frac{\text{length of stick}}{\text{length of crossbar}} \qquad (2.1)$$

then calculations for the gauge shown in Figure 2.10a are

$$k = \frac{76.2 \text{ cm}}{15.2 \text{ cm}} = 5$$

Then, this particular gauge will detect a shrub canopy (or grass, forb, or tree) that is not more than five times its own diameter away from the sample point. Furthermore, that shrub will occupy 1% of the total sample area if a 360° search (to form a large circle) is made from the sample point. If all shrubs that meet this condition are counted, the number of such shrubs will also equal the percentage of total area covered by shrubs at that sample point. To obtain percent cover by species, tally individuals by species as counts are made.

Figure 2.10b illustrates one modification wherein the crossbar is rotatable on a shaft placed through the main stick. This modification can overcome the effect of slope. The bar is rotated in a vertical fashion to become parallel with the slope. Of course, no rotation of the bar is needed if the entire stick can be rotated clockwise to attain the parallelism necessary. The former description of need for a bar to rotate (Fig. 2.10b), while the stick remained stationary, was given because of a mounted eyepiece. If no peephole or sight is mounted, then it is simple to just rotate the whole device. Figure 2.10c illustrates a plant canopy that is more than k times (not counted) its diameter from a sample point, and a plant is not at least k times (counted) its diameter from the sample point. Figure 2.11 illustrates plant canopies to be counted on the basis of the apparent overlap of their canopy with respect to the projected crossbar from a sample point.

A variation in length of the stick and length of its crossbar depends on the particular cover value needed. Ratios of 33:1 cm and 50:1 cm will give basal areas of tree trunks in square feet per acre and square meters per hectare, respectively.

Glass prisms are used to measure basal area covered by tree boles. Sighting through the prism from a sample point will result in a count of boles that are in focus. Otherwise, tree boles may appear to be discontinuous, in which case they are counted if the displacement is less than one-half of the bole diameter (Husch et al. 1982). Distance from the sample point to the tree determines the amount of disalignment of the bole as viewed through the prism. The count is multiplied by a constant to esti-

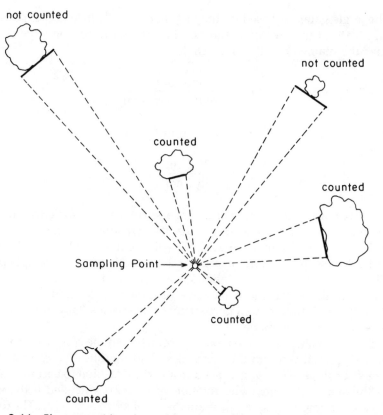

Figure 2.11 Plants considered as within the imaginary plot formed by an angle gauge of Figure 2.10 (Adapted from Grosenbaugh 1952).

mate basal area covered. Decisions on trees to be counted are illustrated in Figure 2.12. Prisms are available in various angle sizes, basal area factors and with plot radius factors.

Other plotless methods include distance measures that are used to estimate mean area per plant, and subsequently, cover can be estimated by species or for total vegetation. First, estimates are made of density using a selected distance measure (from Chapter 5), and for each distance, the plant is measured for diameter. Diameters are used individually to calculate areas of ellipses or circles assumed by plant canopies. Then, an average is obtained for area covered by plant canopies. Density multiplied by the average ground coverage by plant canopies will give the total area covered. The same units used for density, such as individuals per hectare, are used to calculate percent cover. For example, consider an area of one hectare in

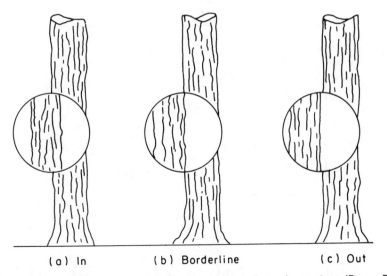

(a) In   (b) Borderline   (c) Out

Figure 2.12 Displacement of tree boles viewed through a prism (From *Forest Mensuration* by Bertram Husch, Charles I. Miller, and Thomas W. Beer, 1982. Reprinted by permission of John Wiley & Sons, Inc.).

size, then

$$\text{Percent plant cover} = \frac{(\text{density})\,(\text{average area of plant})}{10,000} \times 100 \quad (2.2)$$

Consider the need to meet several assumptions to obtain accurate estimates of cover from distance and diameter measures. In particular, consider accuracy of density estimates, accuracy of estimated areas covered by plants, and finally, the product of the two estimates.

2.2 DENSITY

Most density measures use counts from plots or distance measures. Equipment used as a unit to obtain the measure is rather simple. Plot frames are easily made from metal and welded if areas are 1 m^2 or less. Otherwise, boundaries are marked in the field with string or pliable wire held in place by metal pegs. Distance measurements require only a tape measure.

Plots with boundaries of 1, 4, and 10 m on a side form 1, 16, and 100 m^2, respectively, and are used extensively when counts of individual herbs, shrubs, and trees are needed from a forested community. Usually these plots are nested; that is, the 1-m^2 plot is placed in a corner of the 16-m^2 plot, and the latter is placed in a corner of the 100-m^2 plot. Only herbaceous plants are counted by species in the 1-m^2 area, while shrubs are counted

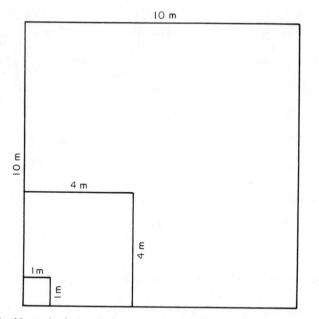

Figure 2.13 Nested plots used to count herbaceous plants, shrubs, and trees, respectively, for 1, 16, and 100 m².

in the total 16-m² area. Trees are counted in all 100 m² of area. Figure 2.13 illustrates the plot arrangement.

No stated rule for a fixed plot size is used to obtain density counts within a given life-form-dominated community. However, the sizes and shapes of plots given above are used extensively to estimate density in grassland (1 m²), shrub (16 m²), and tree (100 m²) communities. The 1-m² plot has been used even in arid shrub communities wherein shrubs make up less than 35% of the total relative cover. However, sample size adequacy for density does depend on the size of the plot used. Therefore, rectangular-shaped plots to estimate density should be considered in any vegetation type so that more variation in plant numbers occur between, rather than within, plots.

Line transects are used to measure the amount of intercept of plants along a straight line. These intercepts are then used to obtain estimates of density. Figure 2.6b illustrates the measure of intercepted length of a shrub or other plant. However, these lengths give unbiased estimates of cover, not density. To obtain estimates of density, the intercepted plant is measured for diameter that is interpreted as widths of canopy rather than a diameter. These widths provide a measure of area and counts within that area. Equations to convert these widths into an estimate are given in Chapter 5.

2.3 BIOMASS

Vegetation biomass is most often determined from plots of given areas. Vegetation in such small areas is usually clipped and weighed to yield an estimate of biomass. Plots of various shapes and sizes are used to obtain the measurement. Shapes include circles, rectangles, and squares, while sizes range from a few square centimeters to several square meters. In general, dense herbaceous vegetation makes small plot areas more efficient, while sparse vegetation allows the use of large areas for measurement of biomass.

Small areas described by plots may yield biased estimates of biomass. If individual plants of bunchgrasses, for example, are so large as to exclude the occurrence of any other plant within the plot boundary, then biomass of the large plants is overestimated, while the biomass of other species is underestimated. Additionally, small plots have larger area:boundary-length ratios.

Table 2.1 gives a comparison of the area:boundary ratios for several plot sizes (area) and three plot shapes. If area is constant (e.g., 1000 cm^2), then plot shape can be changed to increase the amount of area A for a given boundary length B. Note that for a fixed area of 1000 cm^2, the $A:B$ ratio increases from a rectangle, to a square, to a circle (7.14–8.92). This represents a 25% increase in area per unit of boundary length for the circle compared to the rectangle.

Larger areas of plots show less increase in unit area per boundary length. For example, the percentage increase for a circle of 1000 cm^2 compared to an area of 5000 cm^2 is 124%. Yet, an increase in area of twofold (to 10,000 cm^2) yields an increase of 41% in the amount of area per unit of boundary length. The same increases also hold for the square plots of comparable

TABLE 2.1 Boundary Length and Area Relationships as Affected by Plot Size and Shape

Shape (cm)	Boundary Length B (cm)	Area A (cm^2)	$A:B$ (cm)
20 × 50 rectangle	140.0	1,000	7.14
50 × 100 rectangle	300.0	5,000	16.67
50 × 200 rectangle	500.0	10,000	20.00
31.62 × 31.62 square	126.5	1,000	7.91
70.71 × 70.71 square	282.8	5,000	17.68
100 × 100 square	400.0	10,000	25.00
17.84 (radius) circle	112.1	1,000	8.92
39.89 (radius) circle	250.7	5,000	19.94
56.42 (radius) circle	354.5	10,000	28.21

areas (Table 2.1). On the other hand, the rectangular-shaped plots have a ratio increase of 14 and 67%, respectively, for an increase in area from 1000 to 5000 cm² and from 5000 to 10,000 cm².

Based on data in Table 2.1, decisions made to include plants that occur very near or on boundaries of plots are minimized for circular plots. This is true only for plot shape comparisons based on constant area for all shapes. Otherwise, an increase in area or size of a plot can be made for any shape until the area:boundary-length ratio is acceptable. For example, a decision is made to accept a plot, rectangular in shape (50 × k cm), to provide a 10% or less increase in the ratio. Then k is determined to be 180 cm from Figure 2.14. That is, the 10% increase in the ratio is found at a boundary length of 460 cm (area of 9000 cm²). Then, 460 minus 100 (for two sides of 50 cm each) is 360 cm, to be divided equally for the other two sides. A rectangle of dimensions, 50 × 180 cm, will give the desired ratio. Any larger k for the plot will give less than 10% increase in area:boundary-length ratio.

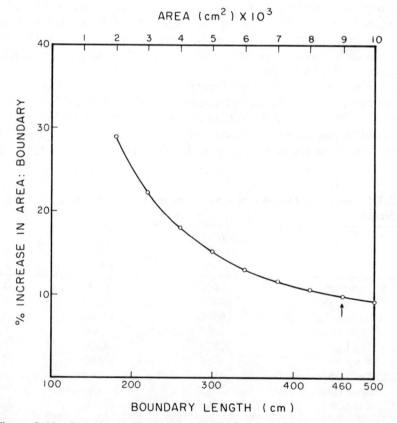

Figure 2.14 Boundary length versus percent increase in area:boundary ratio.

Plots commonly used to obtain weight estimates are illustrated in Figure 2.15. All plots cover 0.5 m² of area. Boundary lengths and area ratios are given in Table 2.1. More boundary decisions are expected to be made for the rectangle plot because more boundary is involved than with either a square or a circle of the same area. Therefore, sampling error is thought to be less for circular-shaped plots. In practice, however, the other two shapes often are as efficient. This is so because of the number of plots needed to obtain an adequate statistical sample. In any case, size and shape of the plot should be determined for each vegetation type. Literature references are useful only if comparisons have been made for the same type of vegetation. Otherwise, the investigator should conduct proper comparisons for each vegetation type to be sampled.

It was previously mentioned that all measurements are estimates of a population parameter, the true value. Then, the term "weight-estimate method" for measurement of biomass in plots is a misuse of the measurement concept. A more accurate description of the process is that of "weight guess." That is, if a plot is used to obtain biomass measures by a double sample of guessing and clipping, both are to be considered as estimates. References are abundant that refer to the clipped and weighed material as "actual." This erroneous concept has led many to discount any estimate other than weight as a valid measure of biomass. Yet, many errors occur

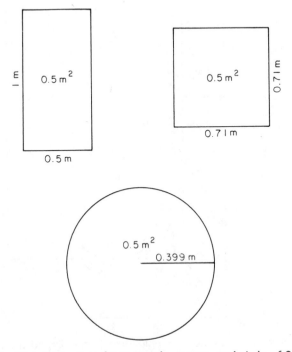

Figure 2.15 Plot dimensions for rectangle, square, and circle of 0.5-m² area.

in clipping and weighing. Errors in weight estimates accumulate as clipping height above ground varies; some clipped material is dropped or not clipped near plot edges, clipping only plants rooted within plots, and bagging material for measurement by a balance.

Experience is needed to obtain guesses and clipped weights to use in regression equations. Training and constant checking by the observer is needed for precision estimates to be obtained. For rangelands, forage production can be simply guessed within practical limits by professionals with many years of experience. Even then, however, such guesses are used only to estimate carrying capacity for livestock. These guesses are seldom accepted in cases where disputes arise because of conflicts with livestock use in competition with uses by wildlife or recreation by humans.

All vegetation biomass is contained within a volume of space, not area alone (Fig. 2.16). That is, the estimate of biomass is based on the projected material above ground level, not on rooted material only. An imaginary bounded volume must be followed in guessing and/or clipping material

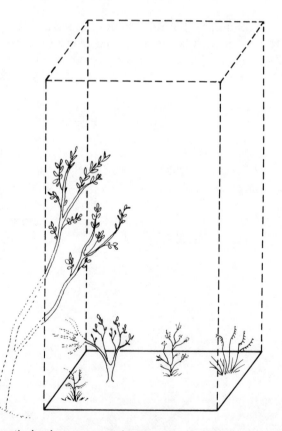

Figure 2.16 Theoretical volume over a plot area. Plants clipped include all material within the volume, not just that rooted within the plot.

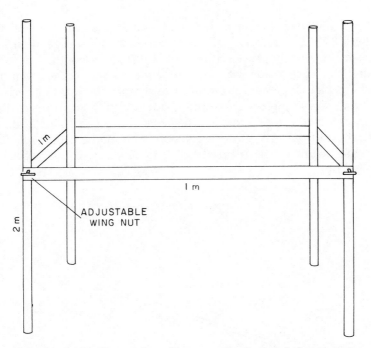

Figure 2.17 Plot used to clip vegetation by height increments.

to estimate biomass. Tall herbaceous vegetation, as well as shrubs, present problems to the observer unless the vertical boundaries are actually extended (Fig. 2.17).

Clever alternatives to plots with rigid boundaries exist. The sickledrat (Fig. 2.18) is, in effect, a circular plot that is formed by a pointed tip. All biomass is clipped within the area inscribed by the tip. The middle rod can be lengthened to include any height of vegetation. Another nonrigid boundary of a circle can also be defined by a string, pipe, or metal rod.

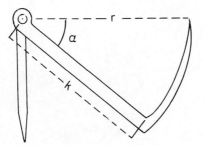

Figure 2.18 Sickledrat. The arm is used to define the plot area only. It is not a cutting edge (From Kennedy. 1972. "The sickledrat: a circular quadrat modification." J. Range Manage. 25: 312.).

Stability is obtained, however, when a metal arm is rotated as illustrated by Figure 2.19. The arm is made to extend above the vegetation height, and a 1-m² area is defined with the arm of 0.56 m in length. Any area can be found by a circle made with simple materials used for the given radius.

Disk- or plate-type meters are used to estimate biomass of herbaceous vegetation. These meters are spring-loaded, and it is assumed that the plate or disk will come to rest at a distance above ground determined by the height and density of the vegetation. The mathematical relation between the height of the plate at rest and biomass of the vegetation is estimated by

$$y = a + bH \qquad\qquad (2.3)$$

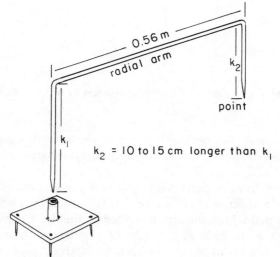

Figure 2.19 A radium arm devise to mark plot boundary on ground surface. Dimensions shown create 1-m² area (From Coleman. 1959. "A useful device for laying out forage production plots." J. Range Manage. 12: 138).

where y represents biomass (kg/ha) and H is the average height (cm) at which the plate stops. The coefficients a and b in the equation are obtained from regression analyses. Obviously, the coefficients are affected by the weight per unit area of the plate. One plate-type meter is shown in Figure 2.20. Some meters (Earle and McGowan 1979) are easier to use than others (Phillips and Clarke 1971). Methods on calibration and modifications of meters are presented in detail in several references (Powell 1974, Castle 1976, Vartha and Metches 1977, Bransby et al. 1978, Michell 1982).

Capacitance meters are constructed with various sizes and shapes of sensing elements. Campbell et al. (1962) were among the first to use this instrument to estimate herbage weights. A capacitance meter with vertical rods is shown in Figure 2.21. Vickery et al. (1980) described a capacitance meter that uses a single rod probe and an electronic system that accumulates the meter readings from a number of locations used as sample points. The problem of obtaining a pattern of rods with a consistently uniform electric field is overcome with a single probe instrument. Otherwise, the range in readings obtained when herbage heights and densities vary may be large with multiprobe instruments. A detailed description of the single probe instrument is found in the work of Vickery et al. (1980).

The single probe capacitance meter is recommended because it is convenient to use. Readings are responses to differences in biomass rather than to differences in moisture content of the vegetation. The capacitance meter is not affected by weather conditions, but plant exudates may cause

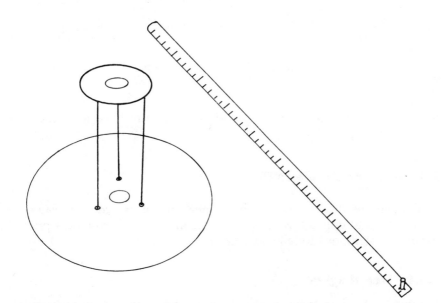

Figure 2.20 Plate meter used to estimate weight of vegetation (drawn from Castle 1976).

Figure 2.21 Capacitance meter.

variable readings. Dead plant materials apparently have little effect on readings.

2.4 TREE MEASUREMENTS

Equipment specifically used to obtain measures of tree characteristics are numerous, and only a few are presented since several references present detailed descriptions for all equipment.

2.4.1 Tree Heights

Heights of standing trees are usually estimated indirectly by instruments called *hypsometers* that use geometric or trigonometric principles. Hyp-

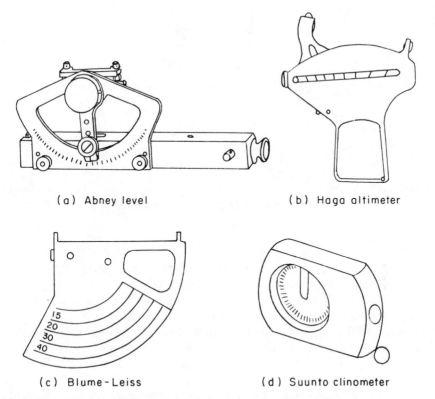

(a) Abney level

(b) Haga altimeter

(c) Blume-Leiss

(d) Suunto clinometer

Figure 2.22   Commonly used hypsometers (From *Forest Mensuration* by Bertram Husch, Charles I. Miller, and Thomas W. Beer, 1982. Reprinted by permission of John Wiley & Sons, Inc.).

someters based on principles of similar triangles include the Christen and Merrit instruments that are sticks with graduated scales (Husch et al. 1982). Some hypsometers require a measurement of the horizontal distance from the observer to the tree. If a slope exists, then corrections for horizontal distance are obtained by trigonometric functions. Some hypsometers require that the distance between the tree and a point be constant, while other hypsometers do not depend on distance to estimate heights.

Hypsometers based on tangents of angles usually provide better estimates of tree heights (Husch et al. 1982). Principles for use of area and angles enable the construction of various hypsometers. These principles are employed in the use of the Abney level, Haga altimeter, Blume-Leiss altimeter, and the Suunto clinometer (Fig. 2.22). These are widely used hypsometers for height and slope measurements. Each type of hypsometer has advantages and disadvantages that depend on topography and density of trees. In general, the measurement is obtained from a position where

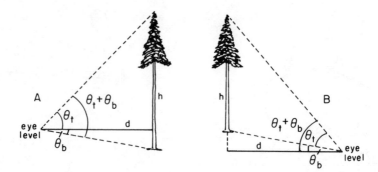

Figure 2.23 Trignometric relations used with hypsometers to estimate tree heights (From *Forest Measurement* by H. C. Belyea, 1931. Reprinted by permission of John Wiley & Sons, Inc.).

both the top and base of the tree can be seen. Line of sight angles to the top and base of the tree are observed (Fig. 2.23).

Tree height is estimated by application of equations using tangents or sines. Note that if the horizontal line of sight intersects a tree between the base and the top, two estimates are made of height, and these two estimates are summed to obtain the total height. Should a measure involve a line of site above or below the base because of slope, then tree height is the difference of the two estimates.

The use of tangents for angles requires a measure of the horizontal distance from a sample point to the tree. Belyea (1931) gives equations to estimate tree heights when angles are measured. If the eye of the observer is above the level of the base of the tree (i.e., up slope from the tree) (Fig. 2.23a), then

$$h = \frac{d[\sin (\Theta_t + \Theta_b)]}{\cos \Theta_t} \tag{2.4}$$

On the other hand, if the eye of the observer is below the level of the base of the tree (i.e., down slope) (Fig. 2.23b)

$$h = \frac{d[\sin (\Theta_t - \Theta_b)]}{\cos \Theta_t} \tag{2.5}$$

If trees are not erect, then the vertical distance from the ground to the top of the tree is measured at right angles to the direction of the lean. Grosenbaugh (1981) described a hypsometer that rotates to provide estimates for the heights of trees that lean. Any hypsometer may not adequately estimate the height of deciduous trees since exact maximum height

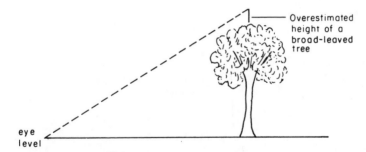

Figure 2.24 Overestimate of height of a broadleaf tree with hypsometers.

may not be seen and is assumed to occur somewhere beyond the first canopy intercept (Fig. 2.24).

2.4.2 Units of Measure for Tree Diameters

Tree diameters are obtained from instruments that range from the simple to the sophisticated devices. These include calipers, tapes, and optical instruments. Diameter measures of trees are usually recorded at breast height (1.4 m), and is commonly referred to as "diameter at breast height" (dbh). A number of field rules exist for obtaining bole diameter measures. All measures should be obtained at the same height on all trees. Then, the diameter of inclined trees is measured at 1.4 m from ground level opposite the direction of inclination of the tree, while the diameter of trees growing on a slope is measured at 1.4 m above-ground on the uphill side. If a tree bole is forked at breast height, then dbh is measured below the point where the bole forks. For boles that fork below breast height, a separate dbh is recorded for each fork. Any tree bole with an irregularity at the base should have dbh measured above where the bole is normal (dn).

Estimates of tree volume are made from "diameter inside bark" (dib) measures. To obtain such a measure, a small portion of the bark must be removed at breast height to measure bark thickness. "Bark gauges" can be used to minimize injury to the tree (Mesavage 1969). Estimates for bark thickness are measured at points on opposite sides of the bole. It follows that dib is obtained by subtraction.

Tree Calipers. Calipers used to estimate diameters of trees are of various sizes and shapes (Figure 2.25). Most calipers use a fixed scaled bar while another piece slides on the bar. The method of obtaining the measurement for diameter depends on the construction of the instrument. To read di-

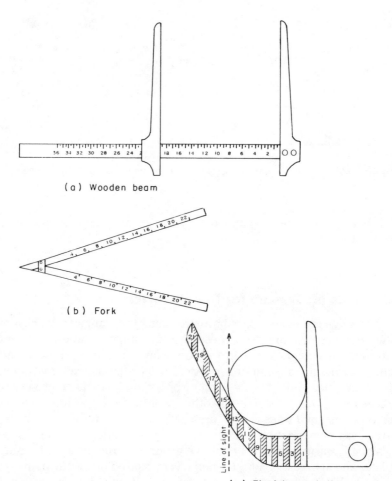

(a) Wooden beam

(b) Fork

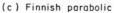

(c) Finnish parabolic

Figure 2.25 Three types of tree calipers (From *Forest Mensuration* by Bertram Husch, Charles I. Miller, and Thomas W. Beer, 1982. Reprinted by permission of John Wiley & Sons, Inc.).

ameter, an average of two diameter measurements is obtained of the longest and shortest diameter for non-circular boles.

Diameter Tape. The diameter of the tree boles can also be measured with a "diameter tape." This tape is like an ordinary measuring tape, but has one side marked in linear units which are converted to a diameter for any given circumference of a circle. The conversion is based on the relationship between circumference and diameter of a circle. That is

$$D = \frac{C}{\pi} \tag{2.6}$$

where

D = bole diameter
C = Circumference
π = 3.14

It follows that a circular shape is assumed for the bole.

Increment Borers. Growth increments in bole diameters are measured by borers. Since the instruments were designed by Swedish foresters, they are often called "Swedish increment borers." A typical increment borer (Fig. 2.26) consists of a hollow auger that is bored into the tree until it intersects the growing center of the tree in a plane perpendicular to the longitudinal axis of the tree. The auger is carefully turned backwards a fraction of a turn to break the wood core, and then the sample core is removed for counting growth rings and measuring the width of each ring.

The age of the tree is estimated from the number of growth rings. Problems in the estimation occur because some trees grow at a differential rate in various seasons, while some growth is affected by meteorological events (precipitation, etc.). "False rings" are present if drought or other unusual

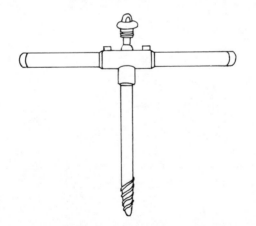

Figure 2.26 Increment borer (From *Forest Mensuration* by Bertram Husch, Charles I. Miller, and Thomas W. Beer, 1982. Reprinted by permission of John Wiley & Sons, Inc.).

events occur. Only through experience in study of material obtained from cores will interpretation be accurate. Such experience must be obtained in the region of interest. Otherwise, increment borers are not suitable to estimate ages of trees.

2.4.3   Tree Crown Cover

The "Moosehorn" is used to estimate the percentage of crown closure or crown cover for stands of trees. Although it was originally developed to estimate volume of timber, Garrison (1949) modified the instrument to obtain crown closure estimates (Fig. 2.27). The dot template shown is transparent and the spots are red so that they contrast with live foliage.

The instrument is a wedge-shaped box with a sighting device and is mounted on a standard surveyor's staff. Mirrors are used to create a vertical projection of the canopy via reflection into the box and onto the dot template. The proper orientation is obtained when images of a marker fall on the center line of a mirror. Line-of-sight is attained when the cross-hair is centered inside the peep-sight ring. Cover is estimated by counting dots on the template covered by the overstory.

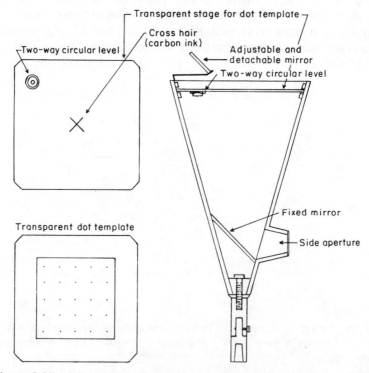

Figure 2.27   A "moosehorn" to estimate crown closure (Garrison 1949).

2.5   BIBLIOGRAPHY

Avery, T. E., and H. E. Burkhart. 1983. *Forest Measurements*. McGraw-Hill, New York, 331 pp.

Belyea, H. C. 1931. *Forest Measurement*. Wiley, New York, 319 pp.

Bransby, D. I., A. G. Matches, and G. P. Krause. 1977. Disc meter for rapid estimation of herbage yield in grazing trials. *Agron. J.* **69**: 393–396.

Buell, M. F., and J. E. Cantlon. 1950. A study of two communities in New Jersey pine barrens and a comparison of methods. *Ecology* **31**: 567–586.

Campbell, A. G., D. S. M. Phillips, and E. D. O'Reilly. 1962. An electronic instrument for pasture yield estimation. *J. Br. Grassl. Soc.* **17**: 89–100.

Castle, M. E. 1976. A simple disc instrument for estimating herbage yield. *J. Br. Grassl. Soc.* **31**: 37–40.

Coleman, S. H. 1959. A useful devise for laying out forage production plots. *J. Range Manage.* **12**: 138–139.

Cooper, C. F. 1957. The variable plot method for estimating shrub density. *J. Range Manage.* **10**: 111–115.

Curtis, R. O., and D. Bruce. 1968. Tree heights without a tape. *J. Forestry* **66**: 60–61.

Earle, D. F., and A. A. McGowan. 1979. Evaluation and calibration of an automated rising plate meter for estimating dry matter yield of pasture. *Aust. J. Exp. Agric. Anim. Husb.* **19**: 337–343.

Garrison, G. A. 1949. Uses and modifications for the "moosehorn" crown closure estimator. *J Forestry* **47**: 733–735.

Grosenbaugh, L. R. 1952. Plotless timber estimates—new, fast, easy. *J. Forestry* **50**: 32–37.

———. 1981. Measuring trees that lean, fork, crook, or sweep. *J. Forestry* **79**: 89–92.

Holmes, C. W. 1974. The Massey grass meter. *Dairy Farming Annual*, pp. 27–30.

Hovind, H. J., and C. E. Rieck. 1970. *Basal Area and Point Sampling: Interpretation and Application*. Wisconsin Conservation Technical Bulletin 23, 55 pp.

Husch, B., C. I. Miller, and T. W. Beers. 1982. *Forest Mensuration*. Wiley, New York, 402 pp.

Kennedy, R. K. 1972. The sickledrat: A Circular quadrat modification use in grassland studies. *J Range Manage.* **25**: 312–313.

Meeuwig, R. O. 1977. Slope-compensating stick-type angle gauge. *South. J. Appl. Forestry* **1**(3): 26.

Mesavage, C. 1969. New Barr and Stroud dendrometer, Model FP 15. *J Forestry* **67**: 40–41.

Michell, P. 1982. Value of a rising-plate meter for estimating herbage mass of grazed perennial ryegrass-white clover swards. *Grass Forage Sci.* **37**: 81–87.

Michell, P., and R. V. Large. 1983. The estimation of herbage mass of perennial ryegrass swards: A comparative evaluation of a rising-plate meter and a single probe capacitance meter calibrated at and above ground level. *Grass Forage Sci.* **38:** 295–299.

Morrison, R. G., and G. A. Yarranton. 1970. An instrument for rapid and precise point sampling of vegetation. *Can. J. Botany* **48:** 293–297.

Neal, D. L., R. L. Hubbard, and C. Eugene Conrad. 1969. A magnetic point frame. *J. Range Manage.* **22:** 202–203.

Nomoto, T. 1975. New type of grass meter for pasture yield estimation. *Japan Agric. Res. Quart.* **9:** 165–170.

Phillips, D. S. M., and S. E. Clarke. 1971. The calibration of a weighted disc against pasture dry matter yield. *Proc. N. Z. Grassl. Assoc.* **33:** 68–75.

Powell, T. L. 1974. Evaluation of a weighted disc meter for pasture yield estimation on intensively stocked dairy pasture. *N. Z. J. Exp. Agric.* **2:** 237–241.

Vartha, E. W., and A. G. Metches. 1977. Use of a weighted-disc measure as an aid in sampling the herbage yield on tall fescue pastures grazed by cattle. *Agron. J.* **69:** 888–890.

Vickery, P. J., I. L. Bennett, and G. R. Nicol. 1980. An improved electronic capacitance meter for estimating herbage mass. *Grass Forage Sci.* **35:** 247–252.

Wheeler, P. R. 1962. Penta prism caliper for upper-stem diameter measurements. *J. Forestry* **60:** 877–878.

3 STATISTICAL CONCEPTS FOR FIELD SAMPLING

The purpose of this chapter is to introduce fundamental statistical concepts and to develop a basis for their use in field vegetation measurements. This chapter does not address statistics as a discipline in detail, as the subject is presented in many books. Instead, emphasis is placed on the underlying principles of statistics, an understanding of which is important to obtain and assess vegetation measurements in an objective manner. Statistical procedures are, of course, very effective ways to summarize and evaluate data.

The statistical approach to making field measurements provides an objective method of obtaining and evaluating vegetation data. Therefore, any methodology that does not incorporate statistical criteria should be used with caution in vegetation measurements. Individual chapters on vegetation measurements will emphasize principles to assist in the selection of a methodology that meets minimum statistical criteria.

Application of statistics in field measurements of vegetation characteristics may be divided into three steps:

1. *Design of the Experiment.* An efficient sampling plan will lead to a reduction in unnecessary data collection. Any sample design should include, beforehand, the kind of statistical analysis to be conducted on the vegetation data collected. It may not always be possible to specify the exact details of all analyses. But at least some general form of analysis, such as analysis of variance or a *t*-test, should be in mind before any measurement is made.

2. *Reduction of Data for Estimation Purposes.* Data reduction usually results in a description of a characteristic of the sample. Such a description is referred to as a "statistic" and may include such estimates as the mean (average) or variance (spread) of the sample data.

3. *Examination of Significance.* Information may be gained if hypotheses are tested. For example, we may question whether the population mean differs statistically from one vegetation type to another. Such a mean value might be total standing biomass or any other vegetation characteristic. Significance, or the lack thereof, should be determined from a randomly drawn sample of the vegetation characteristic to be measured and by testing an appropriately constructed hypothesis about the sampled population.

3.1 CHARACTERIZATION OF DATA

Vegetation data can be presented in one of several ways, including a list, class intervals, or summary form such as an average. A list of data values is more easily read and interpreted if data are arranged from the smallest value to the largest value or in order of the largest value to the smallest value. Class intervals may be used to classify data according to a given interval width. Then, the number of times a value occurs in each class interval can be used to calculate the proportion of values that occur within each class interval. This proportion is referred to as the "relative frequency" of the data. Relative frequencies of data values indicate probabilities of occurrences for the interval values, and this distribution is referred to as a "frequency distribution for the data values." For example, the probability of occurrence, or frequency distribution, of a particular face of a die is $\frac{1}{6}$ and happens to be the same for all faces. These probabilities ($\frac{1}{6}$, $\frac{1}{6}$, $\frac{1}{6}$, $\frac{1}{6}$, $\frac{1}{6}$, $\frac{1}{6}$) indicate the frequency distribution of each numerical value on the six faces of a die. That is the numerical values 1 through 6.

The biomass data in Table 3.1 is a population of possible values in grams per square meter, which has a frequency distribution. This distribution is presented as a histogram (Fig. 3.1). If the same data in Table 3.1 are considered as a subset of a larger population, the frequency distribution shown

TABLE 3.1 Grams of Green Plant Material per Square Meter[a]

(1)	(11)	(21)	(31)	(41)
28	42[b]	38	53	32
(2)	(12)	(22)	(32)	(42)
39	50[b]	49	45	57
(3)	(13)	(23)	(33)	(43)
22[b]	46[b]	54	43	37
(4)	(14)	(24)	(34)	(44)
39	44	45[b]	37	46
(5)	(15)	(25)	(35)	(45)
48	12	44	71	45[b]
(6)	(16)	(26)	(36)	(46)
61	29[b]	31	35	26
(7)	(17)	(27)	(37)	(47)
58	41	37	30	49
(8)	(18)	(28)	(38)	(48)
52	66	43	26	39[b]
(9)	(19)	(29)	(39)	(49)
45	40	50	39	38
(10)	(20)	(30)	(40)	(50)
48[b]	33	39	14	27[b]

[a]Assuming that a plot of 5 × 10 m is the population. Numbers in parentheses indicate plot number. A sample of 10 plots is drawn at random to estimate population parameters.
[b]Drawn at random to be measured.

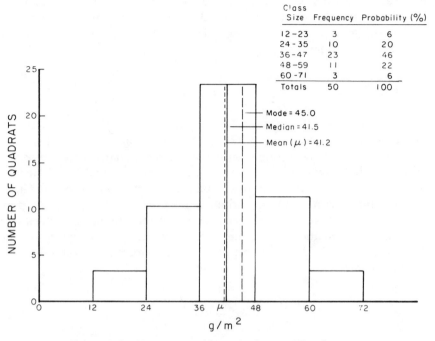

Class Size	Frequency	Probability (%)
12 – 23	3	6
24 – 35	10	20
36 – 47	23	46
48 – 59	11	22
60 – 71	3	6
Totals	50	100

Figure 3.1   Histogram of biomass from a 50-m² area.

in Figure 3.1 will be the frequency distribution of a sample or of a sub-population.

The histogram in Figure 3.1 describes the central location and dispersion parameters of the frequency distribution. The location of these data occurs somewhere between the values of 12 and 72; most observation values occur between 36 and 48. The location of data can be summarized by various measures of central tendency and include the arithmetic mean, the mode, and the median, which are discussed later.

The spread or dispersion of the data is measured by the range in data values, the quartile range, the mean deviation of individual values from the mean, and the standard deviation. These measures of location and dispersion are used to describe sample data or the entire population of values. When these measures are applied to populations, they are referred to as "parameters" but are defined as "statistics" when applied to sample data or a subpopulation data set.

3.1.1   Measures of Data Location

The center of a data distribution, like that shown in Figure 3.1, can be measured in several ways. The center of a distribution of population values is referred to as a "location parameter," which describes the central tendency and is calculated by adding data values together and then dividing this sum by the number of values used in the summation. The process in steps is straightforward. The sample data (Table 3.1) are used to estimate the "arithmetic mean" and

$$\sum_{i=1}^{n} X_i = 393 \text{ g}/10 \text{ m}^2 \tag{3.1}$$

The Greek letter Σ represents addition or summation, and the index $i = 1$ to n indicates that the summation extends over all data values, where n represents the number of data values for the variable X and $n = 10$ for the example. The arithmetic mean of a population is represented by the Greek letter μ, and $\overline{X}$ represents the arithmetic mean of a sample from the population. It is more convenient to let N denote the number of values that occur in the population and n the number of values in a sample. In the example (Table 3.1), $N = 50$ and $n = 10$. Therefore, from Table 3.1, the sample mean is

$$(\overline{X}) = \frac{\sum_{i=1}^{n} X_i}{n} = \frac{393 \text{ g}/m^2}{10} = 39.3 \text{ g}/m^2 \tag{3.2}$$

and the population mean is

$$\mu = \frac{\sum_{i=1}^{N} X_i}{N} = \frac{2060 \text{ g}/50 \text{ m}^2}{50} = 41.2 \text{ g}/m^2 \tag{3.3}$$

Note that $\overline{X} = 39.3 \text{ g}/m^2$ is an unbiased estimate of $\mu = 41.2 \text{ g}/m^2$ if the sample was obtained by random sampling methods.

The second measure of central tendency is called the "median," which is the middle value of a data set if n or N is odd, or it is the average of the two middle items if n or N is even. Then, the median can be calculated for either a population N or a sample n. The median value m for the sample previously used is determined to be

$$m = \frac{42 + 45}{2} = 43.5 \tag{3.4}$$

This value is obtained by first ordering the sample data from low to high values (g/m²). So, the order is

22, 27, 29, 39, 42, 45, 45, 46, 48, 50

Since $n = 10$, an even number, the two middle values are 42 and 45.

For the population of Table 3.1, the median is

$$M = \frac{41 + 42}{2} = 41.50 \qquad (3.5)$$

The population median could also be found from a vertical line that exactly bisects the area covered by the histogram in Figure 3.1.

The third measure of central tendency is the "mode," which is the value that occurs most often in the frequency distribution. The mode for the population is a value somewhere between 36 and 48 g/m^2 (Fig. 3.1). This range is the modal class. From values given in Table 3.1, the modal value is 45 g/m^2, since this value occurs most often. Both population and sample distributions have modal values.

We have considered general methods for data descriptions by use of a graphical display and by calculation of the arithmetic mean, mode, and median, which are all measures of central tendency. Each of these measures takes on a value that describes or represents the sample data set or the entire population. Because all three measures indicate central tendency or location, the selection of which one to use depends on the following:

1. What is the need of a typical or representative value as dictated by the problem? Is an absolute or a relative value needed, or is a middle value or the most common value required?
2. The frequency distribution of the data also affects the choice to be made. Selection of a location measure depends on whether the data are skewed or symmetrical. Compare the two distributions shown in Figure 3.2. Note that both distributions have approximately the same mode but not the same mean. A normal distribution has a mean, mode, and median of equal values, while nonsymmetrical distributions differ in values for these measures.

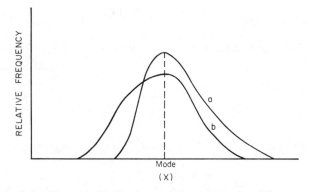

Figure 3.2 Skewed distributions with same mode, different means.

Measures of central tendency provide only a partial description of data. Information on variation or dispersion of the data is also important for fuller description of data. To illustrate the importance of data variation, note that the following sets of numbers have the same mean but different variability as indicated by the range in data values.

Set 1	Set 2
8, 6, 7	3, 16, 3
8, 7, 6	3, 8, 9
$\dfrac{\Sigma X}{n} = \dfrac{42}{6} = 7.0$	$\dfrac{\Sigma X}{n} = \dfrac{42}{6} = 7.0$

Data set 1 ranges from 6 to 8, while set 2 values range from 3 to 16. Therefore, means alone are not equally good descriptors of their respective data sets. The mean of set 2 is said to have more variation than set 1. Some measure of variability or dispersion of the data would add to the information that is conveyed by a value for the mean. We may conclude, then, that a measure of dispersion is also needed to determine the reliability of an estimate of the population mean.

3.1.2   Measures of Dispersion

Measures of dispersion can be obtained from the range in data values, the quartile deviation, the mean deviation, and the standard deviation. These measures may be made from all population values and, like the mean, are called "parameters." They may be made also from sample data values and are referred to as "statistics."

The range is the simplest calculation used to estimate a measure of dispersion and is the difference between the smallest and the largest values of a variable. In the sample data (for sample size of $n \leq 10$) the range r is

$$r = 50 - 22 = 28.0 \tag{3.6}$$

To estimate dispersion or the standard deviation s, the range is divided by 4:

$$s = \frac{r}{4} = \frac{28}{4} = 7.0 \tag{3.7}$$

For the population example, the range R is

$$R = 71 - 12 = 59 \tag{3.8}$$

The standard deviation σ for the population is estimated to be

$$\sigma = \frac{R}{4} = \frac{59}{4} = 14.8$$

The range may be an adequate measure of dispersion, provided the extreme values are not a great distance from the remaining values of the data group. The quartile deviation Q is defined as

$$Q = \frac{(Q_3 - Q_1)}{2} \tag{3.9}$$

where Q_1 is the value of the variable that separates the second quarter (25%) of the data from the first 25% and Q_3 separates the second 25% of the data from the third 25%. For the sample data

$$Q = \frac{46 - 29}{2} = 8.5 \tag{3.10}$$

For the population

$$Q = \frac{48 - 35}{2} = 6.5 \tag{3.11}$$

Since the quartile deviation is one-half of the range of Q_3 and Q_1 values, it is useful as an estimate of dispersion instead of R (or r) when the extremes of the data are widely separated from remaining values. The reliability of this dispersion measure (Q) depends on the concentration of values at the quartiles of the population from which the sample is selected. The presence of discontinuities at quartiles renders the measure unreliable for estimating the standard deviation.

The mean deviation is another measure of dispersion and is the mean of the absolute deviations of all variables from the arithmetic mean. A deviation is obtained by subtracting an individual value of X_i from the mean $\overline{X}$ and using the absolute value of the result in further calculations. Use of the absolute value is necessary since the sum of deviations around a sample mean will always equal zero. Let an individual deviation be represented by a lower case x; then

$$x = X_i - \overline{X} \tag{3.12}$$

where X_i is the ith value and $\overline{X}$ is the sample mean. The mean deviation md is obtained from

$$md = \frac{\Sigma|x|}{n} \tag{3.13}$$

As before, the Σ indicates that the summation is taken over all values in the sample, $|x|$ is the absolute value of the x in Equation (3.12), and n represents the number of observations or measurements in the sample. For the sample

$$
\begin{aligned}
X_1 &= 22 - 39.3 = -17.30 \\
X_2 &= 27 - 39.3 = -12.30 \\
&\quad \cdot \qquad \cdot \qquad \cdot \qquad \cdot \\
&\quad \cdot \qquad \cdot \qquad \cdot \qquad \cdot \\
&\quad \cdot \qquad \cdot \qquad \cdot \qquad \cdot \\
X_9 &= 48 - 39.3 = \quad 8.70 \\
X_{10} &= 50 - 39.3 = \quad 10.70
\end{aligned}
$$

and

$$md = \frac{80.4}{10} = 8.04 \tag{3.14}$$

For the population

$$MD = \frac{446}{50} = 8.92 \tag{3.15}$$

Note that the sample estimate of 8.04 is close to that of the population mean deviation. The mean deviation is seldom used to estimate dispersion because of the requirement to ignore the signs of differences of individual values from the mean. This problem is avoided by using the standard deviation, which is obtained as follows: (1) square the deviation of a value from the mean, (2) sum the squares, (3) divide by N, and (4) take the square root of the quotient from (3). The standard deviation for a population is represented by σ and is obtained from

$$\sigma = \sqrt{\frac{\Sigma_{i=1}^{N} (X_i - \mu)^2}{N}} \tag{3.16}$$

The square of the standard deviation of a population is called the "variance" and is represented by σ^2, where

$$\sigma^2 = \frac{\sum_{i=1}^{N} (X_i - \mu)^2}{N} \tag{3.17}$$

The standard deviation of a sample is represented by s and the sample variance by s^2. The usual case is to compute the standard deviation and variance of a sample in order to make an inference about the population dispersion. When a sample variance is calculated, the divisor $n - 1$ is used in place of n in order to obtain an unbiased estimate of the variance and subsequently the standard deviation.

The equation for the sample standard deviation is

$$s = \sqrt{\frac{\sum_{i=1}^{n} (X_i - \overline{X})^2}{n - 1}} \tag{3.18}$$

and the variance is

$$s^2 = \frac{\sum_{i=1}^{n} (X_i - \overline{X})^2}{n - 1} \tag{3.19}$$

In this case, s is an unbiased estimate of σ, and s^2 is an unbiased estimate of σ^2.

These equations are conceptual in nature and define the standard deviation in an understandable way. If a calculator, which has the mean and variance functions, is not available, use

$$s = \frac{\sum_{i=1}^{N} X_i^2 - \dfrac{[\sum_{i=1}^{n} X_i]^2}{n}}{n - 1} \tag{3.20}$$

The term $\sum X_i^2$ is called the "sum of squares" and the operation calls for each variate X to be squared and the sum of these variates obtained.

Deviations from the mean

$$x_i = X_i - \overline{X} \tag{3.21}$$

also may be presented as standard deviation units. For example, the measurement in row 1, column 1 of Table 3.1 is 28. It deviates from the mean

(41.2) by -13.2 g. This deviation can be expressed in standard units by dividing -13.2 g by σ, which is 11.8 g. That is,

$$\frac{x}{\sigma} = \text{standard deviation units} \tag{3.22}$$

In this case, we have -13.2 g/11.8 g or -1.2 standard deviation units, and the unit of measurement (g/m^2) is eliminated. Therefore, any measurement can be compared to any other measurement on this basis. For example, a measure of grams per square meter can be compared to plant cover in percentage units since standard deviation units eliminate units of measure. It can be verified that all the values for Table 3.1 lie between ± 3 standard deviation units of the mean.

In summary, measures of dispersion provide a basis for comparison of sets of data. The question regarding which measure of dispersion to use for a problem is answered in much the same way as the question regarding which average to use. The selection depends on the nature of the problem, the characteristics of the measure of dispersion, and the type of analysis required.

3.2 PRINCIPLES OF DATA BEHAVIOR

Basic principles of vegetation data behavior are discussed in order to develop an intuitive understanding of variations encountered in the real world. However, variations that are prevalent in measurement data from vegetation systems rarely can be determined exactly. Rather, such variation must be estimated from sample data. Since a vegetation variable does not yield precisely the same value in every sample, it would appear that some degree of certainty needs to be placed on a specified value obtained from a sample process. That is, uncertainty is present in conclusions drawn from sample data collected for a given vegetation characteristic, such as biomass (g/m^2). The presence of uncertainty suggests that relevant factors are not known for sure, but there is still interest in a need to place some level of confidence on a conclusive statement. The variance estimate allows an estimate of confidence to be made as presented below.

3.2.1 Patterns of Data Commonly Observed

In the previous discussion of variations, it was pointed out that data follow certain patterns, and an illustration was graphically presented of the pattern of the data in Figure 3.1. As can be observed, the pattern is symmetrical (bell-shaped) about a certain point. In this case, the mean value of 41.2 g/m^2 occurs at the central part of the curve. Furthermore, a conversion of the data to probabilities did not change the shape of the curve. It is still symmetrical and bell-shaped. This particular type of curve will be closely approximated quite often in nature if data are classified according to frequency of occurrence. Thus, measurements such as plant heights, biomass weights, plant moisture content, and many other variables are approximately "normally distributed." We expect the concentration of values to be at, or near, the central value (location) and to observe an equal distribution (dispersion) on both sides of this central value.

Occasionally we encounter other patterns that do not follow this bell-shaped curve. In other words, they are not normally distributed. Such data occur especially in presence–absence type of data for a plant characteristic. These data give rise to what is referred to as a "point binomial" distribution. These data can take on only one of two possible values (0 or 1) and should be analyzed by binomial methods to estimate parameters.

Another type of data commonly encountered occurs when certain values are clustered together in the population, forming "islands" in the field. This kind of behavior in data is referred to as being "contagious" and forms yet another kind of pattern that differs from the previously mentioned two. Numbers of individual plants in an area, for example, may be clustered under certain conditions with the result that many quadrats may contain few or no individual plants of a species while a few quadrats will contain large numbers of the species. The number of individuals of a species will depend on an environmental factor or set of factors. This kind of data may be a "count binomial," "Poisson," or one of the other contagious distributions presented in Section 3.4. The data should be analyzed by techniques that are designed for the given distribution.

3.2.2 The Normal Distribution

The normal distribution and its parameters are used almost exclusively in applied statistics. The reasons for the wide use of the normal distribution in data analyses is summarized by stating that (1) many populations (weights, heights, etc.) that occur in nature appear to be approximately normally distributed and (2) sampling and subsequent analyses of these populations are easily handled with equations used for assumptions of a normal distribution.

The frequency with which a particular value is expected to occur, if the variable is normally distributed, is calculated by the relation

$$Y = \frac{1}{\sqrt{2\pi}\sigma} e^{-(X-\mu)^2/2\sigma^2} \qquad \text{for} \qquad -\infty < X < +\infty \qquad (3.23)$$

where Y is the frequency of occurrence and X is the variable of interest. This distribution has two unknown parameters. The parameter μ is known as the "mean" (location) of the distribution, while σ^2 is called the "variance" (dispersion) of the distribution. Note that the mean in the example is located under the highest point on the curve in Figure 3.3. This is one characteristic of the parameter μ, and it measures location. The variance σ^2, on the other hand, changes the shape of the curve by spreading the same area over a larger horizontal region of the graph if σ^2 is large, while a small value of σ^2 gives a curve that is horizontally narrow with a higher vertical rise (Fig. 3.3).

There are many normal distributions since there are many possible values for μ and σ^2. As a matter of fact, there is a normal distribution for each possible pair of values that could be assigned to the parameters μ and σ^2. Each population is described when (1) the form (shape) of the frequency curve and (2) the parameters found in the frequency curve in Equation (3.23) are known. Then it follows that the normal distribution of concern is completely described if the mean μ and the variance σ^2 are known. That is, the mean and the variance of a particular normal distribution give all the information necessary about this distribution. Consequently, this is a convenient property to use in data analysis. This does not suggest, how-

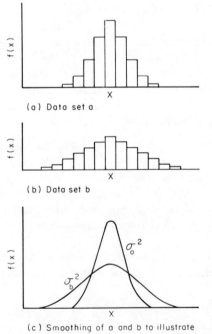

(a) Data set a

(b) Data set b

(c) Smoothing of a and b to illustrate
frequency distribution for continuous data

Figure 3.3 Measures of central tendency of data and variations. Since a distribution is smaller in range than b, the variation in b is greater; c illustrates the principle with continuous data.

ever, that a knowledge of the mean and variance completely describes *any* population, but it is true of populations that follow the normal distribution.

3.2.3 Sample Statistics and the Normal Distribution

In our example, the entire 50-m^2 population was measured. Thus, we know the values of μ and σ^2 for that population. These parameters are rarely known because time and financial limitations allow for measurement of only a few individual items rather than all possible values. Furthermore, it is sometimes necessary to measure a few values because the process of being measured may destroy the material sampled. For example, clipping quadrats to measure above-ground biomass removes the material measured. Therefore, sample estimates are used to estimate the population mean and variance. In review, the sample mean $\overline{X}$ is an estimate of μ, one population parameter, while s^2 is a sample estimate of σ^2, the other population parameter.

The mean and variance of a sample are calculated, respectively, in Equations (3.2) and (3.19). These two functions of sample data are known as "statistics" because they do not contain any unknown parameters (i.e., μ or σ). Other statistics besides $\overline{X}$ and s^2 may be useful to calculate. The mean, median, and mode are all the same numerical value for the normal distribution. This identity is not true for nonnormal distributions (see Fig. 3.2).

The type of relationship expressed by the statistic $\overline{X}$ (or s^2) to the population value is that of a "point estimate"; that is, a specific point (value) is calculated as opposed to an interval estimator. In the latter, a range of values is calculated wherein the true population parameter is expected to occur. For example, the interval for the mean may be expressed by

$$L < \mu < U \tag{3.24}$$

where L represents the lower bound and U the upper bound of μ. If we do not know the values for μ and σ^2, the interval estimation technique allows us to place "confidence limits" on estimates for values of these parameters. That is, a statement can be made as to the certainty (probability) that the calculated intervals will include the population mean and variance, respectively. It was pointed out previously that the normal distribution is a bell-shaped distribution in which the mean, median, and mode all have the same value. It can be proved that if a distance of σ is measured to the right of μ and to the left of μ, then 68.27% of the area under the normal curve will have been included. Any multiple of σ can be used, but usually 1, 2, or 3 is used in practice. From a practical standpoint these results are interpreted as follows. If an item is randomly drawn from a population of values, the probability is 0.6827 (or 68.27%) that its value will be between $\mu - \sigma$ and $\mu + \sigma$. Between $\mu - 2\sigma$ and $\mu + 2\sigma$ includes 0.954 of the

TABLE 3.2   The Possible Values of $\overline{X}$ Given an Example Population of $N = 4$ and a Sample Size of $n = 3$

	Sample No.	Sample Mean
1	31, 34, 37	34
2	31, 34, 40	35
3	31, 37, 40	36
4	34, 37, 40	37

values, and 0.997 of the values lie between $\mu - 3\sigma$ and $\mu + 3\sigma$. Thus, confidence limits can be placed on values of interest.

In most cases, interest is in the mean value (μ) of the population since it is an average value. Therefore, some certainty is needed to ensure that the true mean μ has been included when μ is estimated by the sample mean $\overline{X}$. Suppose that three observations were drawn from a population and the sample mean $\overline{X}$ was calculated. Let us repeat the process and obtain combinations of three observations from a population of only four values: 31, 34, 37, and 40. The means of the possible samples are given in Table 3.2. Each sample gives a mean that may or may not be equal to that of other samples. Nevertheless, the probability of obtaining a given $\overline{X}$ can be calculated if all possible values of $\overline{X}$ are known. In other words, the sampling distribution of $\overline{X}$ is the probability distribution where the possible outcomes are the values that the sample statistic ($\overline{X}$) may assume. Note that possible values of $\overline{X}$ range from 34 to 37, which is a narrower range than values of individuals in a sample. Also note that the distribution of $\overline{X}$ is centered around 35.5, which is the average of $\overline{X}$ values. Call this value $\overline{\overline{X}}$.

Note that 35.5 is also the value of μ, the population mean.

$$\frac{31 + 34 + 37 + 40}{4} = 35.5 \tag{3.25}$$

It is no coincidence that the sampling distribution of $\overline{X}$ is centered about the population mean. The distribution of $\overline{X}$ is not as variable as the individual population samples. Thus, it is seen that $\overline{\overline{X}}$ is centered about μ, and $\sigma_{\overline{x}}^2$ (variance of $\overline{X}$) is smaller than σ^2 (the population variance of X).

The mean of the sampling distribution of $\overline{X}$, $\mu_{\overline{x}}$, is always equal to the population mean μ when simple random sampling is used. This simple equality can be expressed as

$$\mu_{\overline{x}} = \mu \tag{3.26}$$

This relation is very useful in statistics because it shows that the sample mean, based on a simple random sample, neither overestimates nor underestimates the population mean. Therefore, the sample mean $\overline{X}$ is called an "unbiased estimator of the population mean" μ. The term "unbiasedness" refers only to the tendency of sampling errors to cancel out when one considers all possible sample means that can be obtained.

The variance of the population mean σ^2 is

$$\sigma_\mu^2 = \frac{\sigma^2}{N} \tag{3.27}$$

and

$$\sigma_\mu = \frac{\sigma}{\sqrt{N}} \tag{3.28}$$

The variance of a sample mean $\overline{X}$ is

$$s_{\bar{x}}^2 = \frac{s^2}{n} \tag{3.29}$$

The standard error of these means (usually referred to as "standard error" rather than as the "standard deviation of the mean") is calculated as

$$s_{\bar{x}} = \frac{s}{\sqrt{n}} \tag{3.30}$$

Note, then, that the standard error of the sample mean $s_{\bar{x}}$ is always equal to $(\sqrt{n})^{-1}$ times the standard deviation (s) of the sample.

One of the most important theorems in statistics, called the "central limit theorem," states that the sampling distributions of $\overline{X}$ is approximately normal if the simple random sample size is sufficiently large. If the population sampled follows a normal probability distribution, the sampling distributions of $\overline{X}$ is exactly normal for any sample size. This theorem is given here so that the reader may have a better insight as to why the mean is used almost exclusively in placing confidence limits in data analysis. Equation (3.29) shows that the sample size n affects the size in variability of $\overline{X}$, and N affects the size in the population variability of the mean [Eq. (3.27)].

Thus far, the mean and variance have been defined mostly in terms of population parameters. An actual use of these parameters in a more meaningful way is needed. In review, the mean of the sample, denoted by $\overline{X}$, gives some indication of the most likely value that is expected to be encountered, should the sampling of a population be repeated again under the same conditions. The variance of the sample s^2 is the average value of the squared deviations of the individual values from the sample mean. Still another useful statistic is s, the standard deviation of the sample, which is obtained from the square root of the variance.

It was pointed out earlier that the area under a continuous frequency curve represents probability. Thus, to determine a desired probability for a value from a normal distribution, a calculation of the area under the distribution curve is made. However, these laborious calculations can be avoided with the use of a table of areas that apply to the entire family of normal distributions. This table is found in statistical texts under the title "The Standardized Normal Density Distribution." The paperback manual *Tables for Statisticians* (Arkin and Colton 1950) also contains this information. To make full use of this kind of table, a standardization is made for all normal distributions. This is done by transforming the deviation of a value from its mean into standard deviation units as presented earlier. Let Z be the transformed variable calculated as

$$Z = \frac{X - \mu}{\sigma} \tag{3.31}$$

As can be seen, Z is the number of standard deviations that a value X is from its mean μ. From the data in Table 3.1, the example of 53 g is taken from row 1, column 4.

$$Z = \frac{53.0 - 41.2}{11.8} = 1.0 \tag{3.32}$$

That is, the value of 53 g/m^2 found in quadrat 31 of Table 3.1 is one standard deviation more than the average population value of 41.2 g/m^2. From a table of the standardized normal variate Z, it can be seen that the probability that a plot will have between 41.2 and 53.0 g is 0.3413. In other words, 34.13% of the quadrats measured will weigh between 41.2 and 53 g/m^2.

Other intervals of interest can be calculated similarly. Usually, however, interest is in values that lie within plus or minus 1, 2, or 3 standard deviations of their mean. Therefore, probabilities of .682, .954, and .997, respectively, are widely used in analyzing and discussing results.

3.2.4 Confidence Limits

Use of the foregoing information enables the construction of a confidence interval CI around the mean $\overline{X}$. If, for example, a sample of given limits includes the real population mean μ, 95% of the time, the following formula would be used

$$CI = \overline{X} \pm Zs_{\overline{x}} = \overline{X} \pm 2s_{\overline{x}} \tag{3.33}$$

where CI is the confidence interval and $Z = 2$, which will enable 95% of possible confidence intervals to include μ. We usually take only one sample and calculate only one confidence interval.

An example of a confidence interval will illustrate the calculations. Estimates of the mean and variance from the sample obtained from data in Table 3.1 will be used. If

$$\overline{X} = 39.3 \text{ g/m}^2, \quad s^2 = 98.0$$
$$s_{\overline{x}} = 3.1 \text{ g/m}^2, \quad Z = 2.0$$

then for a confidence level of 95%,

$$39.3 - (2)(3.1) < \mu < 39.3 + (2)(3.1) \tag{3.34}$$

or

$$33.1 \text{ g/m}^2 < \mu < 45.5 \text{ g/m}^2 \tag{3.35}$$

Recall that $\mu = 41.2 \text{ g/m}^2$ for the population represented in Table 3.1. Then this particular confidence interval, based on the sample taken, does include μ, and 95% of all possible intervals calculated from samples will also include μ.

3.3 SAMPLE SIZE

Sample size adequacy and its determination is of concern to those involved in vegetation inventories. Sample size is the number of observations made on a measured characteristic of vegetation (e.g., cover or biomass). In the United States, the Office of Surface Mining, Department of Interior, was the first federal agency to require sampling be carried out in a precise manner and with a given level of confidence. Rules and regulations concerning vegetation inventories were established for the Office of Surface Mining (1979), Federal Register Vol. 44, No. 50. Of particular interest for the measurement of vegetation, Section 816.116(b)1,3, states:

> The lower limit for equality is set at 90% of the ground cover and productivity of the reference area with 90% statistical confidence, or with 80% statistical confidence on shrublands, or ground cover and productivity will be at least 90% of the standards set forth in an approved technical guide.

A review of some basic formulas from statistics may clarify the use of certain formulas to estimate an adequate statistical sample size. These formulas are presented in a sequence that leads to appropriate uses of the well-known sample size estimation formula for a univariate, normally distributed vegetation characteristic.

First, formulate a null hypothesis such that H_0: $\mu = k$, where k is an arbitrary constant, and let

$$t = \frac{\overline{X} - \mu}{s_{\bar{x}}}$$ (3.36)

Then

$$t = \frac{\overline{X} - \mu}{s/\sqrt{n}}$$ (3.37)

The value for t is taken from a table, and its value is dependent on the number of observations used to obtain estimates of the sample average $\overline{X}$ and the sample standard deviation s. It further depends on the confidence level desired for using $\overline{X}$ as an estimate of μ, the true population average value. This population average is estimated by $\overline{X}$ of a sample and remains unknown in most vegetation studies.

Then, by rearranging and removing the square-root sign, we obtain

$$n = \frac{t^2 s^2}{(\overline{X} - \mu)^2}$$ (3.38)

where n is the estimated number of observations needed to obtain an estimate of the difference $\overline{X} - \mu$ within a given probability. This probability value is indicated by the selection of t, which is a value from a table of the t-distribution, and the population variance σ^2 is also assumed to be adequately estimated by s^2. Since the estimate of the mean $\overline{X}$ is also obtained from a preliminary sample of an area for vegetation cover or production using a quadrat or other observational units, n is only the first approximation to an adequate sample size. That is, the t-value depends on the size of the preliminary sample [i.e., $n - 1$, which is called the "degrees of freedom" (df)] for estimating $\overline{X}$. Then the solution of n is an iterative process that requires that n observations be obtained and that Equation (3.38) be solved again for $\overline{X}$, s^2, and t until no change in n occurs.

For example, let the value $(\overline{X} - \mu)$ be small so that it is, say, no more than $(.10\overline{X})$. That is, the true difference of the sample mean from the population mean will be no less than 10% of the sample mean $\overline{X}$. It follows that

$$(\overline{X} - \mu) = .10\overline{X}$$ (3.39)

and

$$\mu = (\overline{X})(.90) \qquad (3.40)$$

Since μ is the true population mean and it is unknown, Equation (3.38) is estimated only by substituting Equation (3.40) into Equation (3.38). Assume that a value for s^2 is available from a previous sample. Specifically, it is assumed that vegetation characteristics, such as biomass, are normally distributed with a mean (μ) and variance (σ^2), which are respectively estimated by $\overline{X}$ and s^2. Then

$$n = \frac{t^2 s^2}{[\overline{X} - \overline{X}(.90)]^2} \qquad (3.41)$$

or

$$n = \frac{t^2 s^2}{(.10\overline{X})^2} \qquad (3.42)$$

In general, let $k =$ the proportion or precision that the true difference of the sample mean occurs from the population mean. Then, the general form of the equation for sample adequacy is

$$n = \frac{t^2 s^2}{(k\overline{X})^2} \qquad (3.43)$$

3.3.1 Sample Size for Comparison of Two Means

The assumption made above is that sampling represents an existing condition for vegetation characteristics. If, on the other hand, interest lies mainly in differences that occur between a "before" and "after" perturbation of vegetation mean values, then for detection of a difference level between the two means, $\overline{X}_1$ and $\overline{X}_2$, Equation (3.36) becomes

$$t = \frac{\overline{X}_1 - \overline{X}_2}{s_{\bar{x}_1 - \bar{x}_2}} \qquad (3.44)$$

where

$$s_{\bar{x}_1 - \bar{x}_2} = \left[s^2 \left(\frac{1}{n_1} + \frac{1}{n_2} \right) \right]^{1/2} \qquad (3.45)$$

and $\overline{X}_1$ and $\overline{X}_2$ are before and after perturbation means of vegetation characteristics, respectively, with a common variance s^2. The common variance is estimated from

$$s^2 = \frac{1}{n_1 + n_2} \sum_{i=1}^{n_1} (X_{1i} - \overline{X}_1)^2 + \sum_{i=1}^{n_2} (X_{2i} - \overline{X}_2)^2 \qquad (3.46)$$

Sample 1 is from the area before anything happens to the vegetation, and sample 2 is from the same area after a perturbation occurs to the vegetation; n_1 and n_2 are the respective sample sizes.

If n_1 is assumed to be equal to n_2 in Equation (3.46) and s_1^2 is assumed to be equal to s_2^2, before and after perturbation variances of vegetation measures, respectively, then Equation (3.45) is reduced to

$$s_{\bar{x}_1 - \bar{x}_2} = \left[2s^2 \left(\frac{1}{n} \right) \right]^{1/2} \qquad (3.47)$$

and

$$n = \frac{t^2 2 s^2}{(\overline{X} - \mu)^2} \qquad (3.48)$$

Equation (3.47) explains the "2" in Equation (3.48). The assumptions that $n_1 = n_2$ and $s_1^2 = s_2^2$ may not be justified. Let us suppose that they are. Then, the hypothesis to be tested is

$$H_0: \mu_2 \geq .90 \mu_1 \qquad (3.49)$$

where μ_1 and μ_2 are before and after perturbation vegetation population mean values, respectively. Then, Equation (3.39) specifically becomes

$$\overline{X}_2 = \overline{X}_1 - .10 \overline{X}_1 \qquad (3.50)$$

or

$$\overline{X}_2 = .90 \overline{X}_1 \qquad (3.51)$$

The test of the hypothesis in Equation (3.49) is a one-tailed test, and the standard error of the difference between the means, $\overline{X}_1$ and $\overline{X}_2$ (the before and after perturbation means), is

$$s_{\bar{x}_1 - \bar{x}_2} = \left[\frac{s^2}{n_1} + (.90)^2 \cdot \frac{s^2}{n_2} \right]^{1/2} \qquad (3.52)$$

which reduces to

$$s_{\bar{x}_1 - \bar{x}_2} = \left[s^2 \left(\frac{1}{n_1} + \frac{.81}{n_2} \right) \right]^{1/2} \tag{3.53}$$

This result is based on the assumptions made about $s_1^2 = s_2^2$ and the fact that if X_2 is a function of X_1, then the variance of $X_2 = A^2 \cdot$ (variance of X_1), where $A = 0.90$. If $n_1 = n_2$, then Equation (3.53) further reduces to

$$s_{\bar{x}_1 - \bar{x}_2} = \left[1.81 \frac{s^2}{n} \right]^{1/2} \tag{3.54}$$

If the objective is to obtain an adequate sample that will detect a 10% decrease in the before treatment or perturbation of vegetation characteristics as reflected in the "after treatment" condition of vegetation characteristics, then

$$n = \frac{t^2 (1.81) s^2}{(.10\overline{X}_1)^2} \tag{3.55}$$

if $s_2^2 = s_1^2$, and $\overline{X}_2 = 0.90 \, \overline{X}_1$. Values other than 10% precision of the mean will lead to a different multiplier in the numerator in Equation (3.55). For example, a precision level of 15% will give a value of 1.72. In other words,

$$(1.00 - 0.15)^2 + 1 = 0.72 + 1 = 1.72 \tag{3.56}$$

instead of 1.81 in Equation (3.55).

Sometimes the tabular value Z is substituted for t values. The Z values are obtained from a table of normal deviates and are not affected by n, the sample size. However, the t value is to be used when n is small, say, less than 30, while Z is usually used when n is known to be large. In vegetation cover and production sampling, the use of t in Equation (3.55) provides a good approximation of n, the needed sample size.

3.3.2 Sample Size for Nonnormal Data

It should be noted that n is an adequate estimate of sample size obtained from equations for sample size only because symmetry of the data is assumed. That is, 50% of all observational values lie below the population mean and 50% have values greater than the population mean. If this assumption is wrong and the data are skewed, then the equations are not efficient estimators of n needed for statistical adequacy.

A measure of skewness is beneficial to check for asymmetry in a data set. The formula for skewness can be found in Spiegel (1961). For example, let us use a set of 30 quadrats, 1 m², wherein herbaceous plant production is clipped. Assume values range from 0 to 299 g/m². An estimate of n using 90% confidence and 10% precision is 279. A transformation using the logarithm of the values, recalculating the mean and variance, gives $n = 28$. Although such transformations are valid for nonnormal data in order to obtain symmetry, their use to estimate n may not be recommended because the purpose of transformations is seldom understood. However, forcing the use of normal statistical methods on nonnormal data to estimate n does not yield the desired result, the valid use of data. Transformations are discussed in Section 3.5.

When data do not follow the well-known bell-shaped curve or the preliminary data are very small, another alternative to data transformation may be necessary to estimate the number of samples needed for attaining specified levels of confidence. Substitutions of a probability statement for that of confidence can be made since confidence may imply that an interval of specified distance from the mean is important. On the other hand, there is a way in which a statement can be made concerning the probability that a given percentage of the data mass (concentration) will lie between the lower and upper values of the observation. The result is that an estimate of n can be obtained, that is, the number of observations needed to attain a probability of including a given percentage of possible data values.

Basically, the approach used is that n is a function of α and ω where both of the latter values range between and 0 and 1. That is, the probability is at least α that the probability mass (concentration of values) of F between X_1 and X_n will be at least W (Gokhale 1972). If

$$W = F(X_n) - F(X_1) \tag{3.57}$$

where X_1 and X_n are, respectively, the smallest and largest of the observations, then W (the range in values) is independent of the distribution of the data F. Skewness of data is permitted and the formula for finding $n(\alpha, \omega)$, as given by Gokhale (1972) is accurate for practical purposes when α and ω exceed 0.75. For pre- and postdisturbance (or at two times) inventories of vegetation, the equivalent probability of ω at .80 and .90, respectively, for shrub and grassland characteristics may well suffice. It remains, then, to select an α level that will be accepted and that these former probabilities will have been met by obtaining a random sample of size n. This sample size is obtained by solving

$$A = \ln\left(\frac{10}{1 - \alpha}\right) \tag{3.58}$$

Then n as a function of both ω and α will be the smallest integer that is larger than

$$n = \frac{A \ln[\omega(1 - \alpha)/A] + A - \omega}{1 - \omega + A \ln \omega} \tag{3.59}$$

For example, let $\alpha = .90$ so that at least 90% ($\omega = .90$) of the probability mass will be included if n random observations are made. That is

$$A = \ln \left(\frac{10}{1 - .90}\right) = 4.61 \tag{3.60}$$

and

$$
\begin{aligned}
n &= \frac{4.61 \ln[.90(1 - .90)/4.61] + 4.61 - 0.90}{1 - .90 + 4.61 \ln .90} \\
&= \frac{-18.15 + 3.71}{-0.386} = \frac{-14.44}{-0.386} = 38
\end{aligned} \tag{3.61}
$$

That is, 38 observations are needed to ensure that at least 90% of the data values will occur between the smallest and largest value, 90% of the time. Nothing is implied with regard to the precision of the estimated mean. However, other practical studies have shown that the mean values for cover and production of vegetation change very little as the value of n increases.

3.4 DATA DISTRIBUTIONS

Several data frequency distributions are available for describing the various measures of vegetation characteristics. Most statistical analyses employ the normal distribution; therefore, other frequency distributions can be transformed to obtain normally distributed data. The most commonly used distributions to describe vegetation data are the Poisson, binomial, and log normal.

If data do not fit or are not close to the theoretical distribution known to describe such data, then the data values are not occurring at random. That is, an assignable cause is present, such as soil differences in nutrient availability which cause plant measures to be nonrandom. For example, probability of specific density counts of individual plants occurring within a plot of given area, say, 1 m², can be described by the binomial or Poisson distribution. If this is not true for a given data set, then some environmental characteristic is influencing the numbers of individuals that occur within a 1-m² area. An exploration for the cause should then be made so that the data may be fully explained.

If a data frequency distribution has a single peak, then the measurement reveals a response to a single cause. On the other hand, a bimodal (two humps) is the result of the measurement responding to two or more strata such as environmental differences from one area to another. The data, then, should be categorized into the two groups before analysis is conducted and interpretation of the measurement attempted. If the frequency distribution is relatively flat, with no real peak, the distribution is called a "rectangular distribution." The measurement then indicates that the vegetation characteristic is a response to a mixture of environmental influences and/or management practices. If data cannot be interpreted from the univariate (a single characteristic, as cover of a single species) measurement, then multivariate measurements and analysis will have to be used. To show the relationship among some of the distributions, numerical examples are given.

3.4.1 Bernoulli Distribution

The Bernoulli distribution arises when data are qualitative and only two possible outcomes can occur. For example, the use of a pin to estimate cover of a species provides for either a contact (hit) or no contact (no hit) on a single trial of lowering the pin to ground surface. Similarly, in frequency sampling, a plot either has or does not have an individual of a species within its boundary. These two possible outcomes can be assigned a value of zero (0) for one outcome and a one (1) for the other outcome. Determination of which outcome receives which value is arbitrary and does not affect analysis. For convenience, let the positive (hit or plant occurs within the plot) outcome be one (1) and the absence of a hit or plant from a plot be zero (0) in value. Then, the Bernoulli probability distribution has only two possible values for the random variable x. Therefore, a discrete distribution results and the probability that x has a given outcome is

$$P(x) = p^x q^{1-x} \quad (x = 0, 1) \tag{3.62}$$

where p is the probability that $x = 1$ and q is the probability that $x = 0$ and $q = 1 - p$. Then

$$P(0) = p^0 q^{1-0} = q \tag{3.63}$$

$$P(1) = p^1 q^{1-1} = p \tag{3.64}$$

It should be obvious, then, that the Bernoulli distribution has a single parameter, p. That is, the probability of a success in making a hit on a plant by a pin, or finding a plot which contains the individual plant of interest, is given by p.

The mean of this distribution is

$$\mu = p \tag{3.65}$$

and the variance is

$$\sigma^2 = pq \tag{3.66}$$

and the standard deviation is obviously

$$\sigma = \sqrt{pq} \tag{3.67}$$

Since $p + q = 1.0$, the variance [Eq. (3.66)] is maximum when $p = 0.5$. That is, when the probability of making contact with vegetation from a pin is .50, the variance is .25. As the former probability value increases or decreases, the variance is always reduced in value. Furthermore, the distribution is skewed to the right (tail is to the right) if p is less than .5 and is skewed to the left when p is greater than .5. When p is .5, the distribution is symmetrical.

If n pins are used to count the hits on vegetation, then the mean is estimated from

$$
\begin{aligned}
\bar{p} &= \frac{x}{n} \\
&= \frac{x_1 + x_2 + \cdots + x_n}{n} \\
&= \frac{\Sigma x_i}{n} \quad (x_i = 0 \text{ or } 1)
\end{aligned}
\tag{3.68}
$$

for n pins; $n = 10$ in a point frame. The variance of $\bar{p}$ is

$$V(\bar{p}) = \frac{pq}{n} \tag{3.69}$$

for infinite populations. Assuming that an infinite population is practical for number of possible pins to be placed in a vegetation type, then $\bar{p}$ is an unbiased estimate of p above. That is, a single pin provides a zero (0) or one (1), while more than one pin provides for several opportunities (n, to be exact) to obtain hits or no hits. Then, the probability of a single hit from three pins is [from Eq. (3.62)]

$$P(1) = p^1 q^{3-1} = pq^2$$

If the probability p of a hit on a single pin is 20% ($p = .20$), then

$$P(1) = (.2)(.8)^2 = .13$$

for any one of the three pins to contact vegetation. Since any one pin could be the one making contact, there are three ways such an arrangement could occur:

$$(0,0,1), \quad (0,1,0), \quad (1,0,0)$$

Then

$$P(1) = (3)(.13) = .39$$

That is, the probability is .39 that a hit ($x = 1$) will be made by one pin if three pins are randomly placed. Furthermore

$$P(2) = p^2 q^1 = (.2)^2(.8) = .032$$

But, any two pins of the three can have a hit

$$(0,1,1), \quad (1,0,1), \quad (1,1,0)$$

or three possible arrangements. So

$$P(2) = 3(.032) = .096$$

Then the probability is 9.6% that at least two pins will make vegetation contact if the probability p is equal to 20% that a hit will be made on a single randomly placed pin. To ease computations, use combinatorial formulas. Let

$$\binom{n}{x} = \frac{n!}{x!(n-x)!} \tag{3.70}$$

where ! is the factorial symbol, which means that $n! = n(n-1)(n-2) \cdots (1)$, define $0! = 1$, and $x = 0,1,2, \ldots, n$. Then for the last example

$$\binom{3}{2} = \frac{3 \cdot 2 \cdot 1}{2!(3-2)!} = \frac{6}{2} = 3$$

Equation (3.62) becomes modified to account for n observations made, each of which can take on a 0 or 1. Then

$$P(x) = \binom{n}{x} p^x q^{n-x} \tag{3.71}$$

This probability function is called the "binomial probability distribution." In fact, other sequences of Bernoulli trials (0 or 1) lead to either a geometric,

Pascal, or negative binomial distribution. This knowledge helps to explain why these distributions are often referred to in data analyses of vegetation measurements. The relationships are:

Binomial variate, x = number of successes in n observations.

Geometric variate, n = number of observations up to and including the first success.

Pascal variate, n = number of observations up to and including the xth success.

Negative binomial variate, x = number of failures before the xth success

All observations (trials in some literature) are based on the Bernoulli parameter p and x = 0 or 1 at each observation point, whether the point is a pin, a plot, a line, or another sample unit. The binomial was empirically derived above. The geometric probability distribution is given by

$$P(n) = pq^{n-1} \qquad (3.72)$$

where n = 1 is the number of pins observed before a hit is made and the hit gives n observations. If five pins are observed before a hit is recorded and if p = .20 as before

$$P(5) = (.20)(.80)^4 = .082$$

The probability is 8.2% that five pins are needed before the first hit is made if the Bernoulli parameter p = .20; that is, there is a 20% chance that an individual pin will contact vegetation. The mean of the distribution is

$$\bar{n} = \frac{1}{p} \qquad (3.73)$$

where $\bar{n}$ is the average number of observations needed to obtain the first hit or occupied plot. The variance is given by

$$v(n) = \frac{q}{p^2} \qquad (3.74)$$

The Pascal probability distribution is given by

$$P(n) = \binom{n-1}{n-x} p^x q^{n-x} \qquad (3.75)$$

where n is the number of observations up to and including the xth success.

For instance, if interest lies in calculation of the probability that five pins are observed and two are hits (xth success = the second hit), then

$$P(5) = \binom{4}{3} (.20)^2(.80)^3 = \frac{4!}{3!1!} (.10) = 4(.02) = .08$$

or the probability is 8% that at least two pins will have hits and the second hit is on the fifth pin, given the Bernoulli parameter $p = .20$. In other words, sample until two hits are observed, and note n (the total number of pins observed) to obtain the two hits. The number of occupied quadrats follow the same analysis.

The mean number of observations needed before the xth success is observed is given by

$$\bar{n} = \frac{x}{p} \tag{3.76}$$

and the variance is

$$V(n) = \frac{xq}{p^2} \tag{3.77}$$

The negative binomial probability distribution results when interest in knowing the probability of observing n sample units (pins or plots), wherein the measurement is equal to zero (0) before the xth success occurs. In other words, how many absences or no-hits occur before the xth success? That is

$$P(x) = \binom{x + y - 1}{y} p^x q^y \tag{3.78}$$

where $P(x)$ is the probability that exactly $x + y$ observations must be made to produce x successes, y represents the number of failures to make a hit or find a species in a plot, x is the number of successes in making a contact (hit) or finding the plant, and p and q are as defined previously. Then, if two successes ($x = 2$) and three failures ($y = 3$) occurred

$$P(3) = \frac{4!}{3!1!} (.20)^2(.80)^3 = \frac{4}{1} (.04)(.51) = 0.08$$

The mean is

$$\bar{y} = \frac{xq}{p} \tag{3.79}$$

where $\bar{y}$ is the mean number of failures before the xth success is noted. The variance is

$$V(y) = \frac{xq}{p^2} \qquad (3.80)$$

Note that the mean and variance are the same as those of the Pascal distribution. The probability is very low that three pins will miss before two hits are made when $p = .20$ for each pin. Only the binomial will be discussed further.

3.4.2 Binomial Distribution

The binomial distribution results from the Bernoulli distribution with the parameter p being used repeatedly in a sample size of n to estimate the probabilities of $x = 0,1,2, \ldots , n$ hits by pins, plots occupied by a species, and so on. Then, the binomial variate x represents the number of successes encountered in n Bernoulli trials (repeated measurements for a value of 0 or 1). The probability distribution is given in Equation (3.71).

The mean number of hits or occupied plots expected from n observations is given by

$$\bar{x} = np \qquad (3.81)$$

where n is the number of observations and p is the Bernoulli parameter p. The variance is

$$v(x) = npq \qquad (3.82)$$

The Bernoulli parameter p is estimated by

$$p = \frac{x}{n} \qquad (3.83)$$

where x is the number of hits on plants or number of plots occupied by a plant of interest. Since $x = 0$ or 1, $\Sigma x = x$. If $np > 5$ and if $0.1 < p < 0.9$, then the normal distribution provides a good approximation to x. When $np > 25$, the approximation is good, regardless of the value of p. So data analysis should follow that used for normally distributed data if these conditions are met.

The binomial variate x can be approximated by the Poisson variate with mean of np when $p < .1$. Note the following conditions that are useful as a guide to determining which distribution a data set may follow if the data are discrete and based on the Bernoulli parameter p:

Binomial	variance < mean
Negative binomial	variance > mean
Poisson	variance = mean

For cover estimates obtained by points, the sample adequacy equation (Cochran 1977) is

$$n = \frac{t^2 pq}{(kp)^2} \tag{3.84}$$

where t is the value from the t-table associated with the probability (confidence level) desired, p is the estimate of cover as a proportion, $q = 1 - p$ as before, and k is the precision level with which we wish to estimate the mean cover proportion p. Note that as p approaches 0.5, and q approaches 0.5,

$$n = \frac{t^2}{k^2} \tag{3.85}$$

Then, if $k = 10\%$ of the mean is desired and $t = 1.96$ to obtain a 95% confidence in the estimate

$$n = \frac{(1.96)^2}{(.10)^2} = 384$$

A total of 384 points are needed if the cover of vegetation is near 50%. From binomial tables with the same confidence, it is noted that 250 points will detect a mean cover of 44–56% for a true cover of 50%. For $n = 100$, a range of 40–60% is obtained for the same confidence and precision levels. The sample sizes are for number of randomly located pins.

For points located randomly along transect lines, use Equation (3.43) to estimate the number of transect lines needed from

$$n = \frac{t^2 s^2}{(k\bar{x})^2}$$

where s^2 and $\bar{x}$ are obtained from transect totals. The results from Equation (3.84) dictate the number of points per transect. The optimum number of points per transects (or frames) and the number of transects should be obtained by methods presented as a two-stage sampling process.

The binomial distribution can be used as a discrete distribution to fit a data set containing any measure of number (density, quadrats, etc.) or of proportion, such as cover or fraction of quadrats occupied. However, as n becomes larger and p smaller, the Poisson distribution is an approximation to the binomial distribution. Use t-tests to test hypotheses about p. (See Section 3.6 for the forms of t-tests.)

3.4.3 Poisson Distribution

The Poisson distribution is called the "rare-events" distribution and is a special case of the binomial distribution. That is, the probability p of an event is very small, but there are some occurrences found in a large sample (size n). If n is at least 50 and np is less than 5, then it is a rare event. For example, the observational unit is small relative to the total area. Either a plot or a pin qualifies; the plot could be 1 m^2 or 10 m^2 as long as it is small relative to the total area to be sampled.

Plant density could qualify as a Poisson variate if the number of individuals is low relative to the possible number that could grow in the area. Then, density may be 10 or 100 per unit area as long as the value is low relative to the maximum possible. As density increases to the maximum possible, the values of density per plot will approach the frequency distribution described by the binomial distribution. The Poisson probability distribution is given by

$$P(x) = \frac{d^x e^{-d}}{x!} \tag{3.86}$$

where $x = 0,1,2, \ldots , d - 1, d$ (density) $=$ numbers/unit area, $e = 2.718$, and $x! = (x)(x - 1) \cdots (1)$ as before. Note that d could be the number of hits of occupied quadrats as long as np is small because p is assumed to be small $(P < .1)$.

The mean and the variance of this distribution are equal and are estimated from density d (see Chapter 5) where

$$\text{Mean} = d = \frac{\text{mean number of individuals}}{\text{unit area or time}}$$

The variance of d is

$$v(d) = \frac{d}{n} \tag{3.87}$$

The mean d is the sample mean estimated as usual by $\Sigma \, x/n$. The sample variance, however, is also d, as previously stated. Again, the Poisson variate

x is the limiting form of the binomial variate x as n becomes large, p approaches zero, and np approaches d.

The Poisson variate x may be approximated by the normal variate with mean and variance equal to d if $d > 9$. If the average density d is 3 individuals/m^2, then the probability of finding at least two individuals is

$$P(2) = \frac{3^2 e^{-3}}{(2)(1)} = \frac{.45}{2} = .22$$

from Equation (3.86). The probability of no-hits or having no plots containing the species when $d = 3$ is

$$P(0) = \frac{3^0 e^{-3}}{0!} = .05$$

To estimate necessary size of sample needed, use the form for the normal distribution by converting the Poisson variate to the normalized form if $d < 9$. That is, let

$$y = \frac{x - d}{\sqrt{d}} \tag{3.88}$$

where y will be normally distributed. Equation (3.88) should be recognized as a standardized variable from the relation

$$x_i = \frac{x_i - \mu}{\sigma} \tag{3.89}$$

Then, use the mean and variance of y in the sample adequacy equation for normally distributed variates [Eq. (3.43)]. See Section 3.6 for tests of some hypotheses with t-tests.

3.4.4 Lognormal Distribution

The lognormal distribution describes the probability distribution for logarithm of data values. In the normal distribution, the mean and median are equal in value. The median m and the mean μ are related by $m = e^\mu$ or $\mu = \log m$ in the lognormal distribution. The probability for the lognormal variate is

$$P(x) = \frac{1}{x\sigma(2\pi)^{1/2}} \exp \left\{ \frac{-[\log(x/m)]^2}{2\sigma^2} \right\} \tag{3.90}$$

where x is the variate of interest with values equal to or greater than zero, m is the median value, and other symbols are as described previously. If all values for x are converted to their log values, then these log values follow the normal distribution. The mean μ and variance σ_L^2 of the distribution are respectively

$$\mu = m \exp\left(\frac{\sigma^2}{2}\right) \tag{3.91}$$

and

$$\sigma_L^2 = m^2 e^{\sigma^2} (e^{\sigma^2} - 1) \tag{3.92}$$

where m is the median and is estimated from

$$m = e^{\bar{x}} \tag{3.93}$$

and $\bar{x}$ is estimated from Equation (3.94).

First, transform the data to be approximated by the normal distribution by taking the log of each x; then the mean of the transformed distribution is

$$\bar{x} = \frac{1}{n} \sum_{i=1}^{n} \log x_i \tag{3.94}$$

and the variance is

$$s^2 = \frac{\Sigma \left[\log(x_i - \bar{x})\right]^2}{n - 1} \tag{3.95}$$

The lognormal distribution occurs in vegetation data when the data have a variance that is proportional to the mean. This results when the measurement, say, density in one time period, is proportional to the initial amount present. Also, the biomass of plants at a given time depends on the amount initially present. As time progresses and biomass or density is measured, the mean and variance are correlated over vegetation strata, treatments, time intervals, and so on. Therefore, such data are approximated by the lognormal distribution.

3.5 TRANSFORMATIONS

Transformations of data are made for several reasons. They can allow normal distribution theory to be used for analysis rather than less understood distributions. If a wide range of values exists in the variance from one vegetation strata to another, for example, then a proper transformation of the data may be made to stabilize the variance. That is, the transformation is made to obtain homogeneous variances across topography, strata, or other sources of variation. In effect, the scale of measurements is changed in order to make analysis of variance valid (Bartlett 1947). Then, if the variance changes as the mean changes, the variance will be stabilized by the appropriate change of measurement scale.

The mean and variance are independent in normally distributed data. That is, the correlation is zero between these two parameters as topography, time, or other sources of variation are encountered. Often, such independence between the parameters does not exist, so a transformation is in order. Note that the means and variances within all distributions, except those of the normal, are related. For example, as previously given:

Distribution	Mean	Variance
Binomial	np	npq
Poisson	d	d
Log normal	$me^{(1/2)\sigma^2}$	$m^2 e^{\sigma^2}(e^{\sigma^2} - 1)$
Normal	μ	σ^2

Obviously, the variance of the normal distribution does not contain the term for the mean. However, the latter distribution allows measurements to have values from $-\infty$ to $+\infty$, while the others allow values from zero (0) to $+\infty$ as integers. Thus, transformations may be preferable only when data analysis is used to test hypotheses and such a test assumes homogeneous variances and independently normally distributed errors for the measurements.

The form of the change in variance with the mean must be known or assumed to select the type of transformation needed. Any transformation should yield:

1. A variance that is not affected by changes in the mean as environmental conditions change.
2. A variate that is normally distributed.
3. A variate for which an arithmetic average is an efficient estimate of the true mean for any group of measurements.
4. A new variate that provides for effects (strata, etc.) to be linear and additive.

Transformations provided below make these provisions. See Bartlett (1947) for a full discussion. See Appendix 3.A at the end of this chapter for basis of transformations.

3.5.1 The Square-Root Transformation

Density data consist of whole numbers that often follow a Poisson distribution. The mean is equal to the variance for this distribution, and from

Equation (A.5) from Appendix 3.A we obtain the square root as the transformation necessary to form a new variate with an independent mean and variance. The transformation is not exact but, rather, is approximate since Equation (A.4) is not an equality. In cases where very small numbers are involved (from 2 to 10 for mean values), Bartlett (1947) recommended the use of $\sqrt{x + 0.5}$ instead of $\sqrt{x}$ for square-root transformation. The revision helps especially when zeros occur in the data. Then, the new variate is

$$y = \sqrt{x + 0.5} \qquad (3.96)$$

Malone (1968) used the transformation $\sqrt{x + 0.5}$ to analyze biomass data for peak standing crop of herbaceous shoots of rye grass. He found that the transformed variates fit the normal distribution, whereas the original data differed significantly from normality. In this case, the data were not from a Poisson distribution because data were, in reality, continuous in values. However, when small values (<10) are obtained in a measurement, the square-root transformation will assist in obtaining symmetry in data distribution since skewdness may exist in such data. The estimate of the arithmetic mean is not efficient because the approximation of the original distribution is not based on the Poisson distribution. Any further analysis of such data is only a first approximation to a more exact analysis provided when the exact distribution of the original data is known.

Heady and Van Dyne (1965) used the square-root transformation on biomass composition obtained from point samples placed on clipped herbage. They used regression analyses to predict percentage of species by weight from percentage of contact points determined by the laboratory method described in Chapter 6. Transformation of the data did not provide as satisfactory results as did the raw data. The purpose of the transformation was different for this study and that by Malone (1968). Malone wanted data symmetry so that mean peak standing crop could be predicted adequately. Heady and Van Dyne wanted an equation that best predicted biomass by species from percentage composition of species.

3.5.2 The Logarithmic Transformation

The logarithmic transformation should be used as an approximation to a distribution whereby the mean and variance are proportionally related as in the binomial and lognormal distributions. The log transform applies since

$$\log(XY) = \log x + \log y \qquad (3.97)$$

That is, if the binomial mean and variance are considered, then

$$np = k(npq) \qquad (3.98)$$

and $k = (q)^{-1} = (p - 1)^{-1}$ and from Equation (A.4), $\log(np)$ results, which suggests that data follow a lognormal distribution. The new variate is

$$y = \log x \tag{3.99}$$

where x represents the number of hits by pins or number of occupied plots, and so on.

This distribution applies particularly to measurements of growth in numbers (density) and biomass. These data tend to increase in geometric ratios. If the response of plants behave according to the amount of available nutrients or water, for example, then plant response is not likely to be normally distributed. This response occurs because the limiting factor affects the amount of area that can be occupied by plants. That is, the number of plants and the amount of biomass are restricted by the number of available nutrients in a specific area.

3.5.3   Angular Transformation

The most widely used angular transformation is the inverse sine of the square root. The new variate is

$$y = \sin^{-1} \sqrt{x} \tag{3.100}$$

where x is the percentage measurement such as cover. The inverse sine is measured in degrees and defines an angle of rotation of the data point. Discontinuity often exists in the data, and a useful correction is to add $\frac{1}{4}$ to zero (0) data and leave other integers unchanged (Bartlett 1947). Change n to $n - \frac{1}{4}$ for further analyses of the new variate. Then, convert the new variate to be a measure in radians instead of degrees for analysis. Another useful revision is $\sqrt{x + 1}$ to smooth out the effects of 0 values. However, if 100% does occur, the 1 should not be added.

3.6   FORMS OF THE t-TEST

For comparison of two populations, before and after a disturbance such as surface mining, some information on the t-test is needed. This test is the most powerful test used between two populations if assumptions are met. In practice, one needs to select the proper form of the t-test based on assumptions made about variance σ^2 and size of sample n.

For paired observations X, Y

$$t = \frac{\overline{X} - \overline{Y}}{s_{\overline{d}}} = \frac{\overline{d}}{s_{\overline{d}}} \tag{3.101}$$

$$s_{\bar{d}}^2 = \frac{1}{n}(s_1^2 + s_2^2 - 2s_{12})$$

$$= \frac{1}{n} \frac{\Sigma\, d_i^2 - (\Sigma\, d_i)^2/n}{n-1} \tag{3.102}$$

where

$$d_i = X_i - Y_i$$

$$s_1^2 = \frac{SSX}{n-1}$$

$$s_2^2 = \frac{SSY}{n-1}$$

where SSX and SSY are sums of squares for X and Y, respectively, and

$$s_{12} = \frac{SCP}{n-1}$$

where SCP is the sum of cross-products between X and Y.

For nonpaired observations and $\sigma_2^2 \neq \sigma_1^2$, $n_1 \neq n_2$, the t-test actually will tolerate considerable inequality between two population variances before appreciable changes occur in probability of errors of types I and II. But, in any case,

$$t' = \frac{\bar{d}}{s_{\bar{d}}} \tag{3.103}$$

which is compared to a corrected t' value by

$$t' = \frac{w_1 t_1 + w_2 t_2}{w_1 + w_2} \tag{3.104}$$

where t_i are tabular t values at a chosen probability level with $n_i - 1$ degrees of freedom and

$$w_i = s_{\bar{x}_i}^2$$

t' always lies between and t_1 and t_2 and is needed only when d is a borderline case of being significant. If $n_1 = n_2$, but $\sigma_1^2 \neq \sigma_2^2$, then

$$t = \frac{\bar{d}}{s_{\bar{d}}} \quad \text{with } n - 1 \text{ degrees of freedom}$$

where

$$s_{\bar{d}}^2 = \frac{1}{n}(s_1^2 + s_2^2) \tag{3.105}$$

and $n = n_1$ or n_2 since the latter are equal.

If $n_1 \neq n_2$ and $\sigma_1^2 = \sigma_2^2$, then

$$t = \frac{\bar{d}}{s_{\bar{d}}} \quad \text{with } n_1 + n_2 - 2 \text{ degrees of freedom} \tag{3.106}$$

and

$$s_{\bar{d}}^2 = s^2\left(\frac{1}{n_1} + \frac{1}{n_2}\right) = s^2\left(\frac{n_1 + n_2}{n_1 n_2}\right)$$

where

$$s^2 = \frac{SSX + SSY}{n_1 + n_2 - 2}$$

If $n_1 = n_2$ and $\sigma_1^2 = \sigma_2^2$, then

$$t = \frac{\bar{d}}{s_{\bar{d}}} \quad \text{with } 2(n - 1) \text{ degrees of freedom} \tag{3.107}$$

where $n = n_1$ or n_2 since the latter are equal and

$$s_{\bar{d}}^2 = \frac{2s^2}{n}$$

Calculation of s^2 is the same as above.

APPENDIX 3.A

Consider the derivation or origin of data transformations. Suppose that a measurement y is described by the simple model:

$$y_{ij} = \mu + \tau_i + e_{ij} \tag{A.1}$$

where μ is the overall mean, τ_i is the ith major source of variation such as soil texture (say, before the perturbation occurs), and e is the measurement deviation associated with the ith soil and the jth observation within each soil i. It is assumed that $V(y_{ij}) = \sigma^2$, a constant variance for all i and j. If this is not true, then it can be shown that the function

$$g(z) = g\{E(y) + [y - E(y)]\} \tag{A.2}$$

can be used to obtain an appropriate transformation of y into z. $E(y) =$ the mean of the population. Then, by use of a Taylor expansion of Equation (A.2), we obtain the variance,

$$v[g(z)] \approx v(y)[g'(\mu)]^2 \tag{A.3}$$

where $g'(\mu)$ is the first derivative of the function of μ, or the variance and the mean of y is related approximately as

$$v(y) \approx h[E(y)] \tag{A.4}$$

If the transformed variate $g(z)$ is to have a constant variance, then Equation (A.4) is set to a constant by

$$\begin{aligned}
v[g(z)] &= v(y)\{g'[E(y)]\}^2 \\
&= h[E(y)]\{g'[(E(y)]\}^2 \\
&= \text{constant}
\end{aligned}$$

Therefore, $g'[E(y)]$ is proportional to $\{h[E(y)]\}^{-1/2}$ or

$$g(\mu) \propto [h(\mu)]^{-1/2}\, d \tag{A.5}$$

Then, the appropriate transformation is a square root.

To obtain acceptable results from any transformation, an adequate sample must be taken. All too often, small numbers of observations lead to a skewed distribution and unequal variances.

3.7 BIBLIOGRAPHY

Arkin, H. and R. R. Colton. 1950. Tables For Statisticians. Barnes and Noble, New York, 152 pp.

Bartlett, M. S. 1947. The use of transformations. *Biometrics* **3**: 39–52.

Baskerville, G. L. 1972. Use of logarithmic regression in the estimation of plant biomass. *Can. J. Forestry Res.* **2**: 49–53.

Cochran, W. G. 1977. *Sampling Techniques.* Wiley, New York, 428 pp.

Gokhale, D. V. 1972. Formula for sample size required for specified tolerance. *Am. Stat.* **26**(5): 21.

Hastings, N. A. J. and J. B. Peacock. 1975. Statistical Distributions. Wiley, New York, 130 pp.

Heady, H. F., and G. M. Van Dyne. 1965. Prediction of weight composition from point samples on clipped herbage. *J. Range Manage.* **18:** 144–148.

Malone, C. R. 1968. Determination of peak standing crop biomass of herbaceous shoots by the harvest method. *Am Midl. Nat.* **79:** 429–435.

Munro, D. D. 1974. Use of logarithmic regression in the estimation of plant biomass: Discussion. *Can. J. Forestry Res.* **4:** 149.

Spiegel, M. R. 1961. *Theory and Problems of Statistics.* In *Schaum's Outline Series.* McGraw-Hill, New York.

4 FREQUENCY AND COVER

Frequency of a species is the chance of finding the species when a particular size of quadrat is randomly located within a sample area. Determination of frequency is easiest of all quantitative vegetation measurements, but it is the most difficult vegetation characteristic to interpret. It is simply not an absolute measure and depends on shape and size of the quadrat. Nevertheless, the method is widely used by plant ecologists to monitor changes in vegetation because of its simplicity and speed in obtaining the data. Frequency is more efficient for detecting change in vegetation structure than any other measure because of the time needed to obtain other measures. However, measurements of other characteristics are needed in addition to frequency to provide a complete analysis of the nature of the change. While frequency data will indicate a change in species population, they will not pinpoint the vegetation characteristic that has changed. Since plots are used for frequency measurements, all plants are included, woody as well as herbaceous.

Cover and composition of vegetation are also important vegetation characteristics. These are not only criteria for vegetation description, but are also quite sensitive to biotic and edaphic influences. Although the concept of cover is simple, there is no single definition of the term "cover" that would encompass the broad range of interests of plant ecologists. Generally speaking, in plant ecology, "cover" refers to the vertical projection of vegetation, plants or plant parts, humus, or litter onto the ground when viewed from above.

"Composition of vegetation" usually implies a list of plant species that occur in a particular vegetation type. A simple species list does not convey as much information that would be available if one also knew the proportion of each species in a plant community. In early methods of community description, species were assigned "frequency symbols" by inspection (e.g., rare, occasional, and common). Another method involves numerical ratings of species abundance. The latter procedure is described in this chapter. A third approach to measuring composition is to calculate the contribution of each species to the total vegetation cover of a plant community. All species, woody and herbaceous, can be measured for cover and composition, although methods may differ for various lifeforms.

4.1 FREQUENCY

The term "frequency" was derived from Raunkiaer's work in the early 1900s (Raunkiaer 1934) on recording the presence or absence of species in small plots placed in a plant community. Frequency is, therefore, defined as the number of times a species is present in a given number of quadrats of a particular size or at a given number of sample points. It is usually expressed as a percent of the total number of observations. Frequency percentage is also called "frequency index." Frequency, as already mentioned, is one of the most easily obtained quantitative vegetation parameters and describes the distribution of species in a community. Frequency index is a useful tool for comparison of two plant communities and to detect changes in vegetation structure of a plant community over time.

Size and shape of the plot influence results of frequency determination. Results may be further confounded by the species abundance and patterns of plant growth in an otherwise homogeneous vegetation cover. If the plot unit is too small, the likelihood of a species being recorded is small, and the results of sampling indicate a low frequency value for more common species. Less common species may not be recorded at all. On the other hand, if the sample unit is large, results of sampling may show similar frequency for species that have different distributions in a community. Individuals of a species may, in fact, show aggregation. Therefore, results of frequency determination with a rectangular, square, or circular plot of the same area may differ. Because results are dependent on the shape and size of sample units, frequency of species determined by different sizes and shapes of the plots in two communities or in the same community at two different points in time cannot be compared.

An important consideration in frequency determinations is the decision as to what constitutes the presence or absence of a species within a plot. Raunkiaer (ca. 1900) counted a species if a perennating bud of the species was inside the plot. On the other hand, many ecologists consider that a plant must be rooted inside the plot (Romell 1930, Cain and Castro 1959). However, in the case of creeping or matted types of species, the rooting criterion will result in a lower frequency. Therefore, it is necessary to distinguish between two different usages by referring to "rooted frequency" or "shoot frequency" (Greig-Smith 1983). There is a positive correlation between rooted frequency and density for randomly distributed populations. Similarly, there is a positive correlation between shoot frequency and cover (Greig-Smith 1983). However, there is no definite relationship between frequency and density for nonrandomly distributed plant populations, which makes it difficult to interpret frequency data (Aberdeen 1957, Greig-Smith 1983).

4.1.1 Minimal Area and Frequency

Optimum size of a plot for vegetation measurement has always intrigued plant ecologists. Qualitative ecologists emphasize sampling for recurring plant assemblages and are interested in a plot size on which the species composition of the community is adequately represented. Such a plot is also referred to as "minimal area." However, minimal area can be determined only in a community that is relatively homogeneous and not fragmented. The requirements of a quantitative description of vegetation usually dictate a large number of observations with a smaller plot size rather than a few large "minimal-area" plots. The large number of plots ensure that the area and, thus, vegetation conditions have been adequately covered in the sample.

The selection of an appropriate plot size for measurement is a subjective decision and is based primarily on the size and spacing of individuals of a species. Some ecologists have tried to reach a plot size compromise by taking a sufficient number of smaller plots so that the sum of the areas of smaller plots is equal to, or greater than, the minimal area. Such an objective is achieved either by placing the smaller plots side by side until the size and shape of the minimal area is attained or by random placement of the plots in the community to be sampled. Minimal area is not suitable for frequency measurement because it would theoretically result in 100% frequency for each species. Therefore, the obvious choice is a smaller-sized plot. Because frequency is a nonabsolute measure and depends on the plot size, one is faced with the question, "What size of plot should I use for frequency sampling?"

4.1.2 Quadrat Size for Frequency

Techniques for determining optimum plot size for a measure of frequency are subjective. This is especially true if one is interested in finding a frequency index for all the species in a community. In this case, a plot of any size will measure some species more adequately than others. Apart from other ecological considerations, the aim should be to sample with accuracy and to include a maximum number of species.

The size of unit for frequency measurement is basically a function of plant size and species richness in a unit area. Experimental data have shown that up to 10 species per sampling unit can be counted conveniently. Therefore, the size of the sample unit will vary, within limits, and will depend on the type of vegetation sampled. Raunkiaer employed a circular plot with an area of 0.1 m², but Cain and Castro (1959) suggested the following empirical sizes:

Moss layer	0.01–0.1 m²
Herb layer	1–2 m²

Tall herbs and low shrubs	4 m²
Tall shrubs and low trees	10 m²
Trees	100 m²

Frequencies greater than 95% and less than 5% can result in heavily skewed data distributions. Therefore, the plot size should be such that mean frequencies are larger than 5% but smaller than 95% for all species (Hyder et al. 1965). Curtis and McIntosh (1950) proposed that a plot should be one or two times as large as the mean area per individual of the most common species. Then, randomly distributed species would have 63–86% frequencies for these plot sizes, respectively. Obviously, selection of a single plot size would not be appropriate to measure frequency for the many species of a plant community. The problem is overcome by using a series of plot sizes in a nested configuration that give frequencies between 5 and 95% for a maximum number of species.

Hyder et al. (1965) used the logarithmic relation between density and frequency to calculate the sizes of complementary plots for a nested design. In a randomly distributed population, density d is given by

$$d = -\log_e \left(1 - \frac{p}{100} \right) \tag{4.1}$$

where d is the number of plants per plot area (density) and p is the frequency percentage.

If the density of a species per unit area is known, then its frequency is estimated as

$$p = 100(1 - e^{-ad}) \tag{4.2}$$

where d is density and a is the unit plot area. If we know density d and the unit plot area a, we can estimate frequency for any other given size of plot by the relationship

$$\frac{p_1}{p_2} = \frac{1 - e^{-a_1 d}}{1 - e^{-a_2 d}} \tag{4.3}$$

where p_1 is frequency associated with unit area plot a_1, p_2 is the frequency with another size of plot a_2, and d is density, as before. Conversely, if we know the frequency p_1 for a given size of plot a_1, we should be able to determine the size of plot a_2 for the desired frequency p_2. Equation (4.3) can be written as

$$a_2 = \frac{\log_{10} q_1 - \log_{10} q_2 + .4343 d a_1}{.4343 d a_1} \tag{4.4}$$

where $q_1 = 100 - p_1$
$q_2 = 100 - p_2$
d = density of a species for quadrat (a_1)

Let the area of a small plot a_1 be one unit (e.g., 1 m²), and if the frequency of a species p_1 is 5% (the lower acceptable marginal limit), then from Equation (4.1) we have

$$d = -\log_e (1 - .05) = 0.05 \text{ plants/plot area } a_1 \qquad (4.5)$$

Now that we know the density d for plot a_1 of size 1, in which a species has a frequency of 5%, we can calculate the size of a larger plot a_2 from Equation (4.4), which gives a frequency of 95% (the upper acceptable marginal limit):

$$a_2 = \frac{\log_e 95 - \log_e 5 + (.4343)(.05)(1)}{(.4343)(.05)(1)} = 60 \qquad (4.6)$$

Therefore, if the frequency of a randomly distributed species is 5% by measurement with a small plot, then measurement with a plot 60 times the size of a smaller plot gives a frequency of 95% for the same species. Hyder et al. (1965) computed the theoretical lower limits in the size of complementary large plots for measurement of frequency of a species greater than 85% when its frequency is 5 and 10% in a small plot (Fig. 4.1). The application of this procedure is explained by an example from the work of Hyder et al. (1965).

Hyder et al. (1965) did trial-and-error sampling with quadrat sizes 1652, 523, and 26 cm² to find one that sampled blue grama (*Bouteloua gracilis*) in a frequency range of 63–86%. The plot size of 26 cm² gave an adequate estimate of frequency for blue grama and only four other species. Therefore, to obtain an adequate estimate of frequency for other species, the need of a complementary plot existed. We have already seen that to measure frequency in the range between 5 and 95%, the larger plot should be 60 times the size of the smaller plot. The plot size closest to 60 times 26 cm² was 1652 cm². Hyder et al. used these two complementary plot sizes, and results of their measurements are given in Table 4.1.

It was pointed out above that frequency is not an absolute measure because it is a function of size and shape of the sample unit. However, it can be changed into an absolute measure if the effect of the size of the sample unit is eliminated. This is accomplished by reducing the size of the plot area to a point. A needle lowered at predetermined points over herbaceous cover will either hit or miss a plant part as the needle is lowered. Percentage of hits gives an estimate of the frequency of a species. Then, point data can be described as either frequency or cover. In contrast, presence or absence data using loops or plots can be described only as frequency.

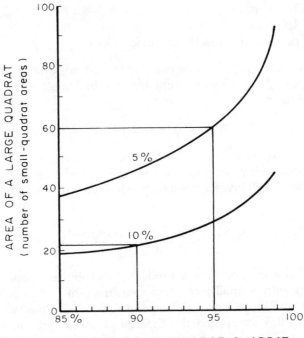

FREQUENCY IN A LARGE QUADRAT

Figure 4.1 Relationship between quadrat size and frequency (From Hyder et al. 1965 "Frequency sampling of the blue grama range." J. Range Manage. 18: 92.).

TABLE 4.1   Frequency Percentages of the Seven Most Common Species Using Quadrats of Appropriate Sizes

Species	Frequencies (%) Using Quadrats Measuring	
	26 cm^2	1652 cm^2
Blue grama [Bouteloua gracilis (H.B.K.) Lag.]	70	100
Buffalo grass [Buchloe dactyloides (Nutt.) Engelm]	25	61
Six weeks fescue (Festuca octoflora Walt.)	16	52
Sun sedge (Carex heliophila Mack.)	14	57
Prickly pear (Opuntia polycantha Haw.)	8	53
Globemallow [Sphaeralcea coccinea (Pursh) Rydb.]	1	20
Common plantain (Plantago purshii Roem. & Schult.)	1	18

From Hyder et al. 1965. Frequency sampling of blue grama range. *J. Range Manage.* **18**: 92.

4.1.3   Frequency Measures

Frequency measurements from a plot, nested plots, or complementary plots are the most common methods for estimation. However, some other methods, such as point sampling and step-point methods developed for cover measurements, yield data that are sometimes used to estimate both species composition and frequencies. The loop method, basically developed to monitor vegetation condition and trend, is also a frequency measurement technique. These measures are described later in this chapter.

Frequency measures can be accomplished either by random or systematic location of plots in the area to be sampled. It is also common practice to select random or systematic sampling points along randomly or systematically placed line transects. However, remember that from the standpoint of statistical analysis, random sampling is preferred over systematic sampling.

Frequency data usually follow the Poisson or point binomial distribution. Therefore, mean and variances are estimated differently from those of normally distributed populations. Frequency, as a proportion, is estimated from a random sample of quadrats as

$$p = \frac{\text{number of plots occupied}}{\text{total number of plots examined}} \tag{4.7}$$

where p is frequency proportion as given in Equation (4.2). Multiplication of p by 100, of course, will convert the estimate to a percentage. Mean and variance of frequency estimates from randomly located sampling units are estimated, respectively, from Equations (4.7) and (4.8) as

$$\text{Variance} = pq \tag{4.8}$$

where p is the frequency percentage and q is the complement $(1 - p)$.

Where a combination of plots and transects is used to obtain an estimate, the variance components of plots σ_{qd}^2 and transects σ_t^2 are estimated from

$$\sigma_{qd}^2 = \frac{k(\Sigma pq)}{n(k - 1)} \tag{4.9}$$

and

$$\sigma_t^2 = \frac{\Sigma p^2 - [(\Sigma p)^2/n]}{n - 1} - \frac{\sigma_{qd}^2}{k} \tag{4.10}$$

where k is the number of plots per transect (constant from transect to transect), n is the number of transects, and p and q are the same as before (Hyder et al. 1965). The subscripts qd denotes that variance is from quadrats (plots).

Optimum sampling depends both on the cost and variance of the frequency data, and there are two components of cost: (1) time involved in locating the observations and (2) recording data at each measurement location. The total time required to record species presence or absence in a plot is negligible compared to the total time required to locate observations. Location of random points is time-consuming. Therefore, to economize the cost of frequency measurements, many ecologists prefer to use a sample design in which plots are located along line transects. Generally, the number of plots is constant from transect to transect.

The optimum number of plots k per transect is given by the relationship

$$k = \sqrt{\frac{\sigma_{qd}^2 C_t}{\sigma_t^2 C_q}} \tag{4.11}$$

The optimum number of transects N is computed as

$$N = \frac{4nV_p}{\left(2\sqrt{\frac{pq}{k-1)}}\right)^2} \tag{4.12}$$

where n is the number of transects already measured and V_p is the variance of the mean frequency percentage of a given species. An efficient sample plan requires computation of k and N for each species. One can then use either the maximum value of k and N or select k and N such that a maximum number or the most important species are adequately measured. The computations of optimum sampling is a lengthy and tedious process. Therefore, for all practical purposes, 25 plots randomly located on 25 randomly located transects should give satisfactory results within a homogeneous plant community.

4.2 COVER

Cover of vegetation is the percentage of ground surface covered by vegetation material. Other definitions exist, but it may be difficult to obtain a measure with respect to the definition. For example, Daubenmire (1959) suggested that cover is an approximation of the area over which a plant exerts its influence on other parts of the ecosystem and is not an estimate of the shaded area on the ground. In other words, plants may be present but only the area of influence may be measured. The inflorescence of plants is often excluded from this consideration since it is relatively evanescent. Seedlings of perennial plants are also sometimes excluded because they may result in an overestimation of cover of perennials in the arid zones immediately following germination. That is, not many of these small plants will continue to provide ground cover. Although the term "plant cover" has a fixed concept, the scope of this definition must be in terms of the interests and needs of the ecologist. Plant cover means different things to different ecologists. The definition of plant cover, to a large degree, depends on the objectives of measurement. Some definitions of cover are:

Vegetation cover—total cover of vegetation on an area.
Crown cover—the canopy of trees including leaves and branches.
Ground cover—cover by plants, litter, and rocks in a vegetation type.
Range plant cover—cover of all plants available to livestock.
Habitat cover—cover of vegetation to protect wildlife.

A characteristic that is common in all the above and any other definitions is that cover is always the vertical projection of vegetation parts, in which one is interested, onto the ground. In addition, habitat cover includes horizontal projection. Cover is expressed as a fraction, percent, or amount of cover on a scale basis.

When cover of a species or plant life-form is expressed as a percent of total vegetation, it is referred to as "relative" cover. Cover is one of the more commonly measured quantities in vegetation sampling. One major advantage of cover as a quantitative measure is that different plant life forms (e.g., mosses, forbs, grasses, shrubs, trees) can all be evaluated in comparable terms. If the vegetation has a distinct layered structure (i.e., shrubs and undergrowth), then depending on the objectives of sampling, the cover of species in each layer is measured separately.

Cover is usually less than total leaf area because many leaves overlap each other. Often there are open spaces in foliage, and in that case, cover will actually be less than total crown or shoot area. Many plants have dissected clumps, and there are gaps in plant canopies. Then dissected clumps should be measured as individual clumps. If there are gaps in the canopy of herbaceous plants, then before measuring cover, each plant may be compressed by hand until the ground surface is not visible through the foliage. The area of dead plant center should be subtracted from the area of the entire clump. It is not possible to compress shrubs and trees. There-fore, the only alternative is to exclude the gaps from measurement. How-ever, consideration should also be given to exclusion of gaps due to shedding of leaves from the branches.

Some ecologists consider small plant gaps as ecologically insignificant and recommend that it may be more meaningful to ignore these gaps. A plant may be considered as an entire clump if individual gaps are less than 2 cm^2. Daubenmire (1968) favored rounding off canopy edges and filling in internal gaps because these gaps may be part of the ecological territory of an individual plant. However, the decision to include or exclude gaps depends on the investigator and objectives of measurement. Whatever decision is made, the measurement procedure should be consistent in order to obtain reproducible results.

The area outline of a plant near the ground surface is referred to as "basal area." This term is widely used in silviculture and range management and refers to the total area of stump surface of trees at breast height (1.4 m) or to area of ground covered by basal parts of grasses, respectively. Basal area measurements are widely used on bunchgrass or tussock vegetation, and measurements are recorded at a height of approximately 2 cm. It is not possible to distinguish basal area in sod-forming, creeping, or spreading types of plants; therefore, only the vertical projection of leaf spread is measured. In the case of straight single-stemmed trees, basal area is measured at breast height, but in the case of trees with multistems or buttresses, basal area is measured at the tree base.

Measurements of basal area are more reliable than aerial cover because foliage cover fluctuates with seasonal changes resulting from climatic fluctuations or other perturbations. The basal area of plants remains fairly constant during a season and may increase or decrease over a period of years. However, in shrubs, forbs, and single-stemmed grasses, basal area is not the best measure because the stem is usually small in comparison to the aerial spread. Basal area measurements have practical application on permanent plots where vegetation changes are to be monitored for several years.

Plant cover can be measured or estimated in several different ways. Quantitative and semiquantitative methods for measures of cover of herbaceous vegetation and woody vegetation are discussed according to techniques used.

4.2.1 Mapping and Charting Methods

Charting methods are of historical interest and are presented mainly from that perspective. However, these methods should be considered in monitoring vegetation changes because they are accurate. A detailed mapping of position and area occupied by species in a plot is referred to as "charting" or "mapping." Plant cover is determined by drawing, to scale, the outline of the crown or basal area of plants on a sheet of graph paper. Charting or mapping may be done with a grid quadrat or a pantograph. A 1-m² quadrat, for example, is divided into 100 dm² by strings. The coordinates are numbered from 1 to 10. The quadrat is placed on the ground, and the area occupied by plants is transferred onto a sheet of graph paper by reading values of coordinates and marking the same points on the graph paper. Different sizes and shapes of quadrats have been used for mapping cover. The choice of appropriate size and shape of quadrat and the size of subsquares depends on the type of vegetation. Since diameter and height measurements are more meaningful measurements for trees, map quadrats are of limited use for estimating shrub and tree cover.

In case vegetation is tall and plant parts are bent by the strings when the quadrat is placed on the ground, the height of the quadrat above ground level may be adjusted by means of adjustable legs. Alternatively, the size of subsquares may be increased by reducing the number of strings. A separate quadrat, the size of one subsquare, with 10, 25, or 100 subsubsquares, may be superimposed on the larger square for mapping cover of vegetation that consists of matted and caespitose species. A partial vegetation map of prairie vegetation near Nebraska, prepared by this method, is shown in Figure 4.2.

The outline of the vegetation can be directly traced onto transparent paper. A glass tracing table, fixed between the legs of a tripod, is used to map cover. An adjustable hole at the top of the tripod is used for viewing

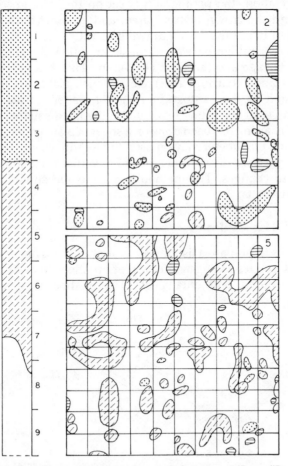

Figure 4.2 Charting plant occurrence along a belt transect (From "Resurvey at end of the great drought" by J. E. Weaver and F. W. Albertson, *Ecological Monographs*, 1943, *13*, 68. Copyright © 1943 by The Ecological Society of America. Reprinted by permission.).

vegetation and tracing its outline on a transparent paper fixed on the tracing table. The table is set about 75 cm above ground and 15 cm below eye position (Booth 1943). This method is used successfully in bunchgrass communities. A box camera, with a glass plate replacing the focusing screen, can also be used to map cover. The camera is mounted on a tripod, and the image of vegetation within a quadrat is focused onto the glass plate. Transparent paper is placed on the glass plate, and the image is traced.

A pantograph may also be used to chart quadrats. A drafter's pantograph (Fig. 4.3) is a device used to reduce the scale of drawings by tracing the original with a pointer connected by arms to a pencil in contact with paper. Hill (1920) was the first to use a pantograph for vegetation mapping. For charting vegetation, the pantograph is set on the ground, the pointer is moved around the crown or basal area, and an exact outline is mapped onto a sheet of paper fixed on a drawing board.

The drafter's pantograph was variously modified by Hill (1920), Pearse et al. (1935), Savage and Jacobson (1935), and Hector and Irvine (1938) to facilitate vegetation charting. The scale at which plants are to be mapped can be adjusted on the pantograph. If graph paper is used for charting, then cover can easily be found by counting the filled squares and fractions of squares. If charting has been done on a sheet of plain paper, a transparent dotted grid or a planimeter may be used to measure cover. The planimeter converts the measure of circumference to area that is compared over time to detect plant changes. The method is reliable and should be useful for permanent quadrats.

4.2.2 Area-List Methods

The first step is to prepare a list of species in the area. Then foliage or basal area cover of species falling within a plot are measured, or ocularly estimated and recorded, onto data sheets. This method is especially suitable

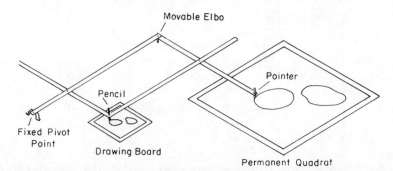

Figure 4.3 Cartographic pantograph (*Aims and Method of Vegetation Ecology*, Mueller-Dombois and Ellenberg, Copyright © 1974 by John Wiley & Sons, Inc. Reprinted by permission of Wiley.).

for vegetation types composed of species that have well-defined clumps, bunches, tufts, or tussocks. Usually, the choice of size and shape of the quadrat depends on the type of vegetation and objectives for measurements. However, 0.5-m^2 (0.5 × 1.0-m) quadrats are commonly used. If plants are small and vegetation is composed of many species, measurements of cover are obtained by serial placement of 0.2 × 0.5-m frames, beginning at the end nearest the investigator. The five placements of subquadrats within the large quadrat constitute one observation for statistical analysis.

Ocular estimates of cover for each species, for area occupied by litter, and for bare ground in a quadrat are recorded to the nearest 1%. Individual observations recorded for the quadrat are then summed. If the sum is less than or greater than 100, the investigator reconsiders the initial estimates and adjusts individual estimates until a total of 100 is obtained.

A number of tools have been developed to facilitate cover measurements for the area-list method. Some of these tools are the listing square (Johnson 1927), diameter ruler (Pearse 1935), densimeter (Culley 1938), and calipers (Murray 1946). The diameter ruler (Fig. 4.4) and calipers have special scales that convert diameter of plants into area. The diameter ruler is used to obtain the diameter of a plant, say, 10 cm; then, by use of πr^2, the area is obtained. That is, $\pi(5)^2 = 78.5$ cm^2. The ruler for area must be marked with the nearest accuracy needed for area estimated.

The listing square (Fig. 4.5) consists of a flat piece of steel 10 cm long and 1.0 cm wide. It is bent in the middle at right angles and the sides are graduated from the center outward in 1.0 or 0.5 cm. The instrument is set closely against foliage, and plant contact with the two arms is read and the two readings are multiplied to obtain the number of square centimeters of cover. The instrument is suitable for basal area measurements where the quarter girth of basal area does not exceed 5 cm. However, instruments with longer arms can be constructed for larger-sized plants.

The densimeter (Fig. 4.6) consists of a steel tape circle mounted on a steel handle. The size of the circle can be adjusted, and the steel tape is calibrated to give measurement of area in 1 × 1-mm (0.01-cm^2) area. The scale is read on the steel handle. Three circles of fixed sizes (0.01, 0.02, and 0.03 cm^2) are attached to the handle to measure cover of seedlings.

Pearse (1935) described the area list method of measuring the cover of range plant populations, particularly forbs and grasses. Pearse used the ruler marked with a special scale from which the area of a circle could be determined by measuring its diameter. Each plant measured is compressed with the hand until ground surface is not visible through the foliage, and the area of the clump is determined with the area ruler. The areas are totaled by species to determine actual and relative covers. Pearse (1935) used 5 × 5-m plots divided by tapes into strips, 1 × 5 m, and measured plants separately in each strip. The procedure yields satisfactory results for bunchgrasses, some forbs with a large basal area, shrubs, and trees.

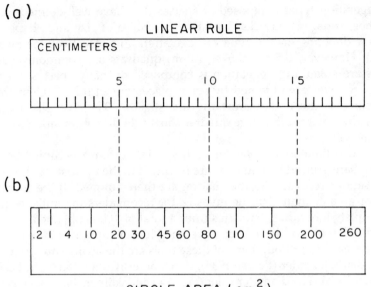

(a)

LINEAR RULE

CENTIMETERS

5 10 15

(b)

.2 I 4 I0 20 30 45 60 80 I I0 I50 200 260

CIRCLE AREA (cm^2)

(c)

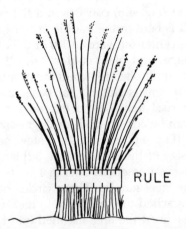

RULE

Figure 4.4 Linear–diameter measure for estimation of basal area of plant (From "An area-list method of measuring range plant populations" by K. Pearse, *Ecology*, 1935, *16*, 573–579. Copyright © 1935 by The Ecological Society of America. Reprinted by permission.).

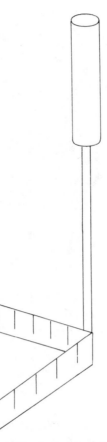

Figure 4.5 Listing square (From "An instrument for list charting" by L. Johnson, *Ecology*, 1927, *8*, 282. Copyright © 1927 by The Ecological Society of America. Reprinted by permission.).

Wright (1972) discussed computer processing of chart quadrat maps for use in plant demographic studies. Charts were made of vegetation in permanent plots on the Jornada Experimental Range in New Mexico. Charts were photographed and then processed through a flying-spot scanner. This procedure coded the charts onto magnetic tape, which was then used for computer input. Analysis of the same quadrat over a period of years yields data on longevity of plant species as well as cover changes.

4.2.3 Photographic Methods

Photographic techniques are rapid, accurate, and objective methods, both for recording data in the field and for subsequent analysis. Techniques for vertical stereophotography are described by numerous authors. Simple

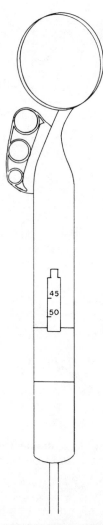

Figure 4.6 Densimeter (From "Densimeter" by M. Culley, *Ecology*, 1938, *19*, 589. Copyright © 1938 by The Ecological Society of America. Reprinted by permission.).

field equipment (Wimbush et al. 1967) consists of a 35-mm single-lens reflex camera with a 28-mm F3.5 wide-angle lens mounted on a rectangular frame with four detachable legs (Fig. 4.7). The frame and legs hold the camera in a downward-facing position, 1.2 m above ground level. The camera mounting consists of a slide with plane screws to permit the camera to be moved 76 mm laterally for taking stereoscopic pairs of photographs. This equipment covers an area of 1 m² (125 × 80 cm) by each stereoscopic pair of photographs. Rectangular (76 × 50-mm) white cards with relevant information regarding the particular quadrat or transect are placed in each

(a)

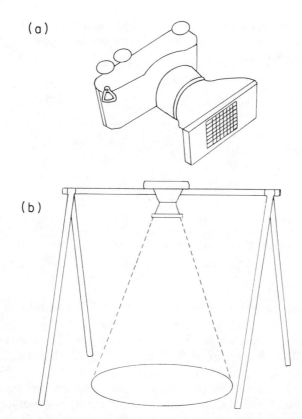

(b)

Figure 4.7 Photographic method for ground cover (From "Color stereophotography for the measurement of vegetation" by D. J. Wimbush et al., *Ecology*, 1967, *48*, 152. Copyright © 1967 by The Ecological Society of America. Reprinted by permission.).

photograph. Wells (1971) modified the technique of Wimbush. He used two 35-mm cameras fired simultaneously to obtain stereophotographs, and Ratcliff and Westfall (1973) used an inexpensive stereo adapter for a single-lens camera to obtain stereo pairs.

Color transparencies are superior to black-and-white photographs because species identification is more accurate (Wimbush et al. 1967). Also, the area of bare ground, litter, and small herbs can be easily separated with color photographs (Table 4.2). Similar results were obtained for standard point procedures. The few discrepancies between the two methods occurred because the photographs measured a larger area. Additional advantages for color transparencies over black-and-white photographs are: the costs of film and processing are less and plant species vigor can be identified.

An excellent study by Wein and Rencz (1976) used color-positive film and the point method to estimate plant cover in vegetation ranging from

TABLE 4.2   Cover Estimates from Color Transparencies and Black-and-White Photographs in Two Areas in the Snowy Mountains Australia, Expressed in Percentage Cover

	Area A		Area B	
Cover type	Black & White	Color	Black & White	Color
Tussock	27	28	0	0
Herbs	50	18	67	60
Shrubs	2	4	0	0
Litter	21	50	9	14
Bare ground	0	0	19	21
Rock	0	0	5	5

From "Color stereophotography for the measurement of vegetation" by D. J. Wimbush et al., *Ecology*, 1967, **48**, 152. Copyright © 1967 by The Ecological Society of America. Reprinted by permission.

polar deserts to meadows in the high Arctic. A 50-m baseline was marked at 1-m intervals at which plant cover was measured as a point. The point frame contained 10 pins located at 10-cm intervals. Quadrat size was adjusted from 50 × 100 cm to 50 × 50 cm to 50 × 25 cm using 2 × 0.4-cm aluminum bars painted in 10-cm divisions. Color-positive film (Ektachrome X) was used in a 35-mm single-lens reflex camera mounted on a tripod to photograph quadrats. The 50 × 50-cm quadrat was photographed with a 35-mm lens from a vertical distance of about 1 m, while a 28-mm lens was used at a similar distance for a 50 × 100-cm quadrat. A binocular microscope was used to estimate cover from the photographs.

Sample sizes required for 95% confidence level, and 20% of the sample mean were calculated for each habitat type and sampling procedure from Equation (3.43). Comparisons of the cover sampling methods from the Wein–Rencz study are given in Table 4.3. There was little variation among quadrat sizes within any one method based on mean total plant cover for the 17 stands. However, sampling time could be reduced by 25% by using the 0.50-m² quadrat rather than the 0.125-m² quadrat.

The photo-quadrat method was the most efficient based on field time only. When laboratory time was considered for the photo-quadrat method, the point method was the most efficient, whereas the line-intercept, photoquadrat, and field-quadrat methods being less efficient, in that order.

Single photographs can be used in open vegetation. In most other vegetation types, however, stereo pairs are preferable. Stereoscopic pairs of photographs can be viewed under a pocket stereoscope for charting or listing. Stereoscopic transparencies can also be viewed directly by transmitted light under a zoom stereoscope. Plant cover can either be charted by projecting transparencies through a pair of prisms onto a glass table or

TABLE 4.3 Mean Cover Component Data (%) and Required Sampling Numbers (N Req.) for Sampling Methods Summarized over Stands Having Greater than 3% Cover of Any Component

| | Point Method Number | | | | | Quadrat Size (m)² | | | | | | Line Intercept (m) | |
| | | | | | | Photo | | | Field | | | | |
	10	20	30	40	50	0.125	0.25	0.50	0.125	0.25	0.50	1	2
Total													
N req.	100	64	52	36	31	81	54	34	113	98	67	76	56
$\overline{X}$	81	79	80	79	80	59	59	57	49	49	49	60	60
Mosses													
N req.	125	81	63	58	46	117	95	64	275	214	178	60	50
$\overline{X}$	60	58	59	59	61	54	55	52	36	37	38	59	60
Vascular Plants													
N req.	108	82	65	53	39	89	60	58	138	95	83	95	93
$\overline{X}$	27	26	25	25	23	12	14	14	8	9	10	11	10
Lichens													
N req.	116	69	55	52	44	74	65	42	126	105	88	88	64
$\overline{X}$	35	34	32	31	30	30	28	28	29	28	27	26	26

Reproduced with permission of the Regents of the University of Colorado from *Arctic and Alpine Research*, Wein and Rencz, Vol. 8, 1976.

may be estimated by point counts. In practice, a 100 × 150-mm image is a convenient size.

Rephotographing of plots for vegetation trend analysis should be done at the stage of plant development that existed for earlier recording. Again, such a method is not suitable when vegetation has several layers except to determine species composition of dominant and, perhaps, subdominant strata. Photographic methods are also subject to inherent errors that result from distortion of plants and parts at the sides of photographs and overestimates of the size of plants and parts standing above the general surface. Both errors are minimized if the height of the camera is large in relation to the height of plants and are not important if the objective is to determine trends in vegetation rather than its actual condition.

Photographs can be used for charting outlines of plants onto transparent paper, and cover is estimated by counting squares or by a planimeter. Cover can also be estimated from photographs by the area-list method with a dot-grid overlay. It is not possible to determine basal area from vertical photographs. However, oblique photographs can be used for measurement of one dimension of the basal spread of a plant. The photographic technique permits a technician with relatively little training to do fieldwork and leave identification and analysis for experienced investigators.

Photographic techniques described so far are for estimating ground vegetation cover. Photographs of tree canopy are taken with a fisheye lens mounted on ordinary cameras. A fisheye photograph is a projection of a hemisphere onto a plane. These photographs provide a relatively accurate method of canopy cover analysis and require less time in the field. Chan et al. (1986) evaluated forest vegetative cover with a computerized analysis of fisheye photographs and found no significant difference in cover estimates obtained by manual and computer analysis of photographs. Although fisheye photography yields reliable estimates, it has not been used because of problems with sunflares and reflectances on photographs.

4.2.4 Intercept Techniques

Intercept techniques, for measurement of cover, use linear measurements of intercepts of lines by plants. These intercepts of the vertical plane of a horizontal line; the number of contacts made by a point with plants; or counting of ocular sightings of plants with a system of cross hairs, grid points, or dot matrix on a transparent surface are all measures of cover. The use of a line for measures of cover by intercept distance is called the "line-transect method." In comparison, a system of cross-hairs, grid points, or dot matrix are grouped under the heading of "point-intercept methods." In fact, both lines and points are plots. The line quadrat is the logical outcome of a two-dimensional quadrat when its size is reduced in one direction until it is reduced to a line. Similarly, a point quadrat is a logical outcome of a two-dimensional quadrat when its size is reduced in both directions until it is reduced to a point.

Line-intercept and point-intercept methods are two of the most popular methods used to estimate cover. These methods are much more efficient than charting methods, and if not more accurate, the results are at least as accurate as those obtained with charting methods. Some studies indicate that higher cover estimates are obtained with a point method than with the line-intercept technique (Johnston 1957, Whitman and Siggeirsson 1954), while others have shown little difference between the two techniques (Heady et al. 1959, Winkworth et al. 1962). The line-intercept method is perhaps most useful for open-grown, woody vegetation. On the other hand, the point-intercept method is more useful for pastures and grasslands. However, the two methods are sometimes combined.

A review of methods of surveying grasslands by Crocker and Tiver (1948) has shown that line-intercept and point-quadrat methods appear to be the only objective methods. The point-quadrat method is preferred because:

1. It provides an objective estimate of cover.
2. It is more expedient to use than other cover methods of equal reliability and objectivity.
3. Randomization and replication is possible.

4. Most species cover estimates are more precise and accurate.
5. Vegetation is less disturbed by the measurement.

Point Intercept. Technically, a point has dimensions associated with area and, therefore, is a plot. The technique for measurement of cover by point quadrats was in use in New Zealand for 8 years before it was first described by Levy and Madden in 1933 (Levy and Madden 1933). Subsequently, it was tested and improved by Goodall (1952, 1953), Winkworth (1955), and Warren-Wilson (1959a,b, 1960, 1963a,b) for use in different vegetation types. The philosophy behind this technique is that if an infinite number of points are placed in a two-dimensional area, exact cover of a plant can be determined by counting the number of points that hit a plant. The technique, in its usual form, is more suited to herbaceous vegetation such as shortgrass prairie and grazed pastures. Specific discussions follow on each method. The point method gives measures of (1) percent ground cover by species, (2) percent cover each species contributes to the total area, (3) relative frequency of species, and (4) percentage that each species contributes to a vegetation type (number of hits per species per number of vegetation hits). This last calculation is related to actual weight if "bulk-for-bulk" species weigh the same.

The point-intercept method can also be used to determine tree canopy cover. In fact, a rifle telescope with a cross-hair can be converted into a point sampler type of periscope to estimate tree canopy cover (Morrison and Yarranton 1970). For reading straight down or upward, a right-angle prism is mounted to the objective lens of the telescope. The rifle scope is mounted on a 3-m-long aluminum beam, held horizontally above ground by a pair of telescoping aluminum tubing at each end (Fig. 2.4). It is usually set up at 1.6 m above ground for convenient viewing and can be moved horizontally along the beam to any position. This enables one to take a number of observations at randomly selected positions along the beam at each placement of the apparatus. The periscope can be rotated by 180° to read ground vegetation cover on the same location as canopy cover.

Grid-Quadrat Frame. Grid-quadrat frames can be of any shape and size. Cross-points of a grid quadrat are, in fact, point quadrats. Vertical interceptions of cross-points with plant parts are considered hits. The angle of viewing introduces error, and two persons viewing at different angles obtain different cover values. This problem can, however, be overcome by having a double grid of cross-points, which greatly reduces the possibility of aiming at different points by different observers. For most grasslands, the small decimeter quadrat shown in Figure 2.8 can be converted into a double grid of 100 cross-points. It can be either used alone or in combination with a square-meter frame.

Stanton (1960) described an ocular point frame to estimate cover in shrub communities. The frame consists of two sets of cross-hairs forming a grid with 25 points, spaced 7.5 cm apart. The frame is supported at convenient heights with metal legs. The two sets of cross-hairs are attached to the frame with the points vertically aligned. The unit is well adapted for measurement of cover in sparse vegetation up to a height of about 1.5 m. Placement of a periscope on a grid system is used to estimate tree canopy cover on the same basis.

Point Frame. A point-frequency frame (Fig. 2.2) is a more practical tool than the grid quadrat and is more frequently used than the latter. It consists of a wooden or a metallic frame with two legs and two cross-arms (Levy and Madden 1933). The two cross-arms have 10 or more perpendicular equidistant holes. Steel rods or wire pins are slid through the holes, and hits are recorded by species. One can either use the same pin for 10 observations or use a separate pin for each hole. However, experience has shown that working with a single pin is slower than having 10 pins. The size of the frame can be designed to suit local vegetation conditions. That is, vegetation height and patterns affect the spacing of pins and height of the frame. Moreover, a periscope mounted on a frame (Fig. 2.8) and readings taken at spaced intervals will yield cover estimates of tree canopies.

The first interception of a pin with a plant shoot part is recorded for foliar cover. Basal area cover is measured by counting only those interceptions which occur with stem parts at ground level. A worked example is given in Table 4.4.

Single Points. A given number of points, distributed individually, can give a more precise estimate of cover (Table 4.5) than if the same number of points are grouped into frames (Blackman 1935, Goodall 1952, Greig-Smith 1957). Single-pin measurements require one-third as many points as required when groups of pins are used. Comparable accuracy (Goodall 1952) and time required is reduced from one-sixth to one-eighth of that required for the point-frame method (Evans and Love 1957). In vegetation types with large-sized or clumped plants, the same plant is intercepted more frequently with more hits per frame. This results in overestimates of cover of such plants. Both the number of pins and the distance between pins affect results. Distance to be used between systematically located pins depends on plant distribution patterns, distance between plants, and size of individual plants (Fisser and Van Dyne 1966). All discussion of single points also applies to the use of a periscope for tree canopy cover measurements.

Angle of Placement. Vertical, horizontal, or inclined placements of pins influence the value of cover estimates. Vertically positioned pins tend to hit flat-bladed species, such as forbs, more often than grasses. Inclined pins tend to favor grasses (Winkworth 1955). Tinney et al. (1937) recommended

TABLE 4.4   Example of Computations of Percent Cover by Point Sampling Methods

Species	Number of Intercepts[a] in 1000 Point Quadrats[a]	Percent Cover[b]	Percent Composition[c]
Blue grama (*Bouteloua gracilis*)	237	23.7	38.4
Western wheatgrass (*Agropyron smithii*)	93	9.3	15.1
Needle-and-thread (*Stipa comata*)	42	4.2	6.8
Buffalograss (*Buchloe dactyloides*)	128	12.8	20.7
Sand dropseed (*Sporobolus cryptandrus*)	27	2.7	4.4
Sedge (*Carex* spp.)	57	5.7	9.2
Three awn (*Aristida* spp.)	36	3.6	5.8
	—	—	—
Total vegetation	620	62.0	100.0
Litter	138	13.8	
Bareground	242	24.2	
Total	1000	100.0	

Data collected near Fort Collins, Colorado.

[a]Sampling was by a point frame with 10 pins. Observations were recorded at 100 randomly selected locations. A pin lowered each time was considered a point quadrat. Data may also be analyzed by considering one point frame (10 pins) as one quadrat. This will give 100 random data points.
[b]Percent cover = (no. of point intercepts/total points) × 100.
[c]Percent cover composition = (no. of point intercepts/total point intercepts of vegetation) × 100.

that pins be inclined at a 45° angle for grasses, and Warren-Wilson (1963b) recommended that an angle of 32.5° be used, which minimized underestimates of leaf area for grasses. Leaf size for trees also will give different results and angles of inclination need to be determined for each forest type when a periscope is used.

Diameter of Pins and Points. Diameter of the pins and points affects the accuracy of cover estimates. Winkworth (1955) determined the effect of point size on heathland vegetation. He concluded that points of finite diameter overestimated percentage cover until pin diameter exceeded the maximum size of gaps in foliage. In this study, even fine pins apparently exceeded the sizes of foliage gaps for these species, and both sizes of pins actually overestimated cover.

TABLE 4.5   Variances for Cover of Individual Species Sampled by 100 Randomly Located Points Compared to 100 Points in 10-Point Frames

Species	10 Frames	100 Points
Sedge		
Carex hebes	35.03[a]	10.92
Australia bluegrass		
Poa caespitosa	35.06	9.54
Purple violet		
Viola betonicifolia	16.57	16.48
Mountain woodruff		
Asperula gunnii	37.83	8.23
Yam-daisy		
Microseris scapigera	21.12	3.61

Goodall. 1952. Point quadrats. *Aust. J. Sci. Res. Series B* 5: 11. Reprinted with permission from CSIRO Editorial and Publishing Unit.

[a]Variances calculated after arc sine transformation of percent cover values.

Pins less than 1.85 mm in diameter are considered impractical for field use because they bend, sway in the wind, and are easily damaged. Therefore, a cross-wire apparatus may be used (Winkworth and Goodall 1962). Comparison tests made among methods indicate that many significantly lower estimates are obtained with the cross-wire. Errors result from finite diameter points, and the magnitude depends upon the leaf size and shape. Errors resulting from the finite diameter of the points are greater for leaves with high perimeter:area ratios (Warren-Wilson 1959b). This error is expressed as a percentage of true foliage area as

$$E = \frac{100d}{lb} (d + l + b) \tag{4.13}$$

where E is the error, d is diameter of point quadrat, l is the length of leaf, and b is the breadth of the leaf. Leaves are considered to have elliptical shapes or to be a combination of ellipses. Percent errors estimated from Equation (4.13) and experimentally determined by comparing areas determined by 2-mm quadrats and by true point quadrats, are similar. Experimental errors range from 13 to 340% and are highest for a narrowleaf species and lowest for a broadleaf species. Error depends more on leaf width than on length.

Error in percentage cover is determined only if an assumption is made about the way in which the foliage is dispersed. Assume that the foliage is randomly dispersed. Then,

$$C = 1 - e^{-r} \tag{4.14}$$

where $C = 0.01 \times$ percentage cover and $r = 0.01 \times$ relative frequency. Use of a pin of a certain diameter increases the relative frequency to an extent that is found in Equation (4.13), provided b and l/b are known. Hence, the increased percentage cover resulting from increased point diameter can be found.

A graph of the theoretical relation of error to leaf breadth is given in Figure 4.8 for pins of 2–4-mm thickness that covers most of the range of

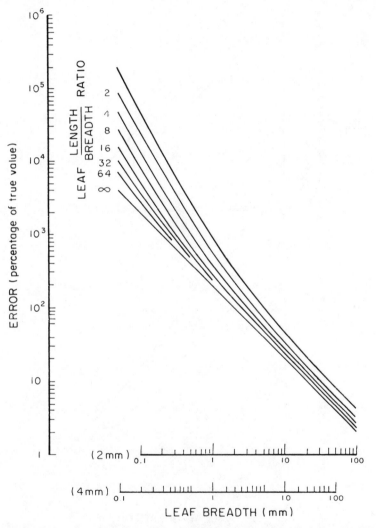

Figure 4.8 Errors in estimation of leaf area for points 2 and 4 mm in diameter. Curves are for leaves of various shapes (length:breadth ratios) (From "Point quadrats errors" by J. Warren-Wilson, *Australian Journal of Botany*, 1963, *11*, 180. Copyright © 1963 by the CSIRO Editorial and Publishing Unit. Reprinted by permission.).

needles used. A separate curve for each leaf length:breadth ratio is given. Ratios are chosen rather than leaf length since the ratio is a measure of leaf shape and varies less than length among leaves of a particular species. The error is doubled for each halving of leaf breadth or doubling of pin diameter for long, narrow leaves, and the error is slightly more than double for round leaves. Expected error was determined from the graph in Figure 4.9 using mean length and breadth for each species. Large leaves were

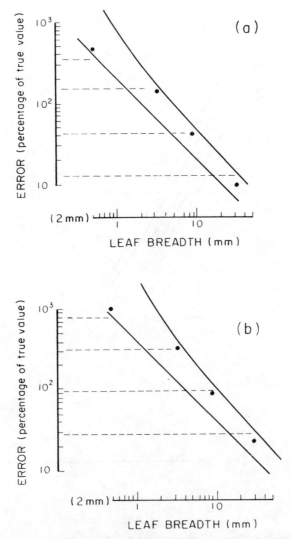

Figure 4.9 Errors in estimates of relative frequency in four species of different leaf size and shape: (a) 2 mm; (b) points of 2 and 4 mm in diameter, respectively (From: Warren-Wilson. "Point quadrats errors." Australian J. Botany 11:180,182. Copyright © 1963 by the CSIRO Editorial and Publishing Unit. Reprinted by permission.).

weighted more than smaller leaves because they affected percentage error more. Errors agreed with those from Figure 4.8. Thus, Equation (4.13), with its assumption of elliptical leaf shape, does provide an adequate estimate of error.

Kemp and Kemp (1956) discussed the analysis of point-quadrat (frame) data from the standpoint of statistical distributions. The moment and maximum likelihood equations for parameters of the beta distribution are

$$\text{Mean } (j) = \frac{v}{v + m} \tag{4.15}$$

and

$$\text{Variance } (k) = \frac{vm}{(v + m)^2(v + m + 1)} \tag{4.16}$$

where j and k are the mean and variance from the beta distribution and v is the number of successes in m trials. The mean number of contacts per frame is nj, where n is the number of frame placements. The variance of number of contacts per frame is

$$v = nj(1 - j) + n(n - 1)k \tag{4.17}$$

if the expected proportion of contacts had this variability from frame to frame. Then, $n(n - 1)k$ measures the variance in excess of that expected on a positive binomial assumption where there is constant expected proportion.

The variance of cover data is usually stabilized with the angular transformation. (See Chapter 3 for information on this transformation.) This transformation is appropriate under either the positive binomial assumption or the hypergeometric type IIA assumption.

The total number of pin observations needed, then, is reduced if the number of pins per frame is decreased. However, the number of locations would probably be increased. The variance of percentage cover value estimated from N_n frames of n pins is

$$V = \frac{j(1 - j)}{N_n n} \left(\frac{v + m + n}{v + m + 1} \right) \times 10^4 \tag{4.18}$$

Patchiness in species cover can be studied if there are at least two pins used per frame so that intraframe variance can be compared with interframe variance. That is, compare results from Equation (4.18) to that from Equation (4.17).

Measurement of Foliage Pattern with Point Quadrats. Warren-Wilson (1959a) used vertical and horizontal point quadrats to analyze the spatial (or pattern) distribution of foliage. It was noted that point quadrats underestimated cover for erect-leaf species and overestimated cover for species with nearly horizontal leaves. The attitude of the leaves of most species varies with environmental conditions. Warren-Wilson suggested that the leaf angle α between the leaf and the horizontal can be expressed as

$$\tan \alpha = \frac{\pi}{2} \left(\frac{F_h}{F_v} \right) \tag{4.19}$$

for flat leaves at a fixed angle where F_h and F_v are the number of contacts with foliage per unit length of point quadrat (contacts per centimeter) set for horizontal and vertical quadrats, respectively. Cover may be estimated as foliage denseness. Foliage denseness F is the total area of foliage per unit volume of space and is estimated as

$$F = \sqrt{\frac{\pi^2}{4} F_h^2 + F_v^2} \tag{4.20}$$

Assumptions with a two-dimensional quadrat are that the leaves have no thickness, face all directions with equal frequencies, and all leaves slope at the same angle within a particular layer. However, errors introduced by violations of these assumptions are not serious. Therefore, F is useful in describing patchiness in foliar cover, while the angle of leaf orientation is useful in interpreting how the patches occur with respect to leaf angle. Then, point angle to obtain cover estimates can be properly selected for each type of species and combination.

If a count is made only for leaf surfaces and not edge contact, then the assumption of foliage thickness is avoided. Most foliage for combinations of plants point in all directions, but the foliage angle in a given layer may differ because of differences found among species or their environment. Separate data should always be kept for species because the assumption of uniform foliage angle results in underestimates of foliage denseness when foliage slopes at different angles. However, Warren-Wilson found that if the divergence in angle between two species of groups was less than 30°, then the error was, at most, 3% for foliage denseness.

Leaf-area index (LAI) is the area of foliage per unit area of ground. A more accurate estimate of cover by LAI could be obtained with quadrats inclined at 32.5° and multiplying the number of contacts per quadrat by the factor 1.1. Warren-Wilson (1963a) found the error could be reduced to 2% by using pins at two inclinations (13° and 52°) and the equation

$$C = 0.23C_{13} + 0.78C_{52} \tag{4.21}$$

Three different angles can be used to reduce the error to 1% but is probably not worth the effort. These errors are shown to be maxima based on the assumption of all foliage slopes at the same angle, which, of course, is a rare case in nature. Preliminary estimates should be obtained from few frames at one angle in a time of 5–20 min. Separate inclined frames should always be oriented in different directions to avoid any bias in the estimates. Warren-Wilson suggested that if both foliage angle and LAI is needed, then vertical, horizontal, and inclined (32.5°) point frames are recommended, and point contacts with vegetation are counted for each centimeter of height. Leaf-area index is estimated within 3% by

$$\text{LAI} = 0.43L_0 + 0.70L_{32.5} + 0.28L_{90} \tag{4.22}$$

where L_0, $L_{32.5}$, and L_{90} are the mean number of pin contacts for pin angles of 0, 32.5, and 90°, respectively. Some species may have foliage evenly distributed across height, while others have mostly basal foliage, and foliage angle of a given species will sometimes vary with height.

Cross-Wire Sighting. Since experimental data have shown that the size of pins affects cover estimates, a cross-wire sighting tube provides a practically dimensionless point-sample (Winkworth and Goodall 1962). For example, a tube 20 cm long with a diameter of 5 cm contains a fine cross-wire at each end. The tube is either held by hand or mounted on a tripod.

Burzlaff (1966) used a telescope attached to a circular disk mounted on a tripod and called it a "focal-point technique." The telescope, from a surveyor's transit or level, is mounted pointing down on a circular board which, in turn, is mounted on a second circular board, both 50 cm in diameter. Cross-hairs in the telescope serve as a point quadrat, and species identification is possible through the lens. This method eliminates bias of point placement.

Goodall (1952) took a sample of 100 quadrats before and after clipping vegetation down to 15–20 cm. He found that the estimate of cover of short species was twice as much after clipping tall vegetation. There was no significant difference in cover estimate of tall vegetation before and after cutting. Short species are likely to be underestimated by this method because of the need to sight through several layers of vegetation.

Line Intercept. The line-intercept method was originally described by Tansley and Chipp (1926). A line transect, sometimes called a "one-dimensional transect," is used for making observations on a line. The method consists of measuring the intercept of each plant under a line, and the line is usually placed on the ground. The intercept of tree species crowns or boles can also be measured with a tape and recorded by species or species groups. Cover of grasses and forbs is measured on the line at ground surface while shrubs, half-shrubs, and trees are measured on the crown spread intercept. Crown spread of tall trees more than 15 m are not easy to measure accurately. Therefore, accuracy of the method depends largely on accuracy of the vertical projection. This difficulty can be overcome to some extent by use of a sighting instrument to project the edge of the crown onto the tape by holding the instrument vertically from the tape to crown edge of the tree.

Borman and Buell (1964) measured tall trees by this method and used a "cover-sight" (Buell and Cantlon 1950) for vertical projection (Fig. 2.1). Cover sight is similar to "moosehorn," except that a single cross-hair is used instead of a dot grid. Another cross-hair is used on the peephole to eliminate parallax problems in sighting. In order to obtain a truly vertical projection, a plumb bob is hung inside the periscope. Jackson and Petty (1973) used an army tank periscope to construct a device for measuring vertical projection of tree crowns. Lindsey (1955) described a "sighting level" for vertical projection of crown outline onto the tape. It consists of a 1.5-m-long stick with a screw mounted into the top end and a carpenter's level mounted 30 cm from the lower end of the stick. The carpenter's level is used to control vertical direction of the stick. The crown outline is a source of large variation. If the tree has a broken canopy within the intercept by the tape, gaps must be excluded from the measurement. Small gaps within the canopy can, however, be ignored. The argument is that these gaps may be part of the ecological territory of an individual.

Canfield (1941) described the line-intercept method for estimating cover of grassland and shrub vegetation. The sample unit was a line transect with length and vertical dimensions only. The intercept of the plants by species, through which a vertical plane of the transect must pass, was measured directly. The line was placed randomly to obtain the estimates. Equipment included some type of line (wire, rope, steel, tape, etc.), two pins for securing the line tightly on either end, and a hammer for driving the pins. Total intercepts of the line by each species are calculated and expressed as a percentage of the total line length to give a direct estimate of percent cover by species.

Parker and Savage (1944) tested the reliability of the line-intercept method in the southern Great Plains in Oklahoma. A 10-m-long steel wire was used for the transect, which was strung tightly at the upper heights of shrubs and anchored by 1-m-long steel pins driven into the ground. The peripheral spread of foliage cover of shrubs and the actual ground cover of all species intercepted by the transect were measured. The shrub cover, which lay within 5 cm of the transect, was considered to be intercepted by the plane. The transect wire was lowered to ground level, and the portions of actual ground cover by species for grasses, sedges, and forbs that lay within 0.5 cm of the transect were measured. Differences in cover at the 1% probability level were found among lines and among observers for perennial grasses but not between replicate measures by the same observer or interactions of the replicates with observers or lines. Hence, the data were reproducible by the same observer but differed among observers.

Hormay (1949) discussed obtaining better records of vegetation changes with the line-intercept method. Accuracy and consistency in the use of the line-intercept method revolves around the determination of the end points of the line. Two characteristics, in particular, must be clearly visualized: (1) the unit of plant measurement and (2) the normal foliar density of the species. The most practical unit of plant measurement is the whole plant, rather than some portion, such as a single stem. If only a portion of the plant were taken for a measurement, bias results because normal interspace distances in the parts of different species also differ. Therefore, the entire plant should be used for the unit of measurement to simplify data collection.

The line transect can be of any length, and choice depends on the type of vegetation being measured. Canfield (1941) recommended a 15-m line for areas with a cover of 5–15% and a 30-m line where the cover is less than 5%. Time to measure a transect is also used to determine the length of the transect. Canfield suggested that the optimum length of a transect is one that can be measured by a team of two persons in about 15 minutes.

The line-intercept method can only be applied to plants (trees, shrubs, grasses) that have a solid crown cover or have a relatively large basal area. Parker and Savage (1944) and Stephenson and Buell (1965) considered tree and shrub branches closer than 10 cm as solid crown. If the canopy is broken, cover of individual components should be summed and recorded as one entry. This facilitates count of individuals for the purpose of density. Therefore, the line intercept is more useful for estimating cover of shrubs and trees in open-grown, woody vegetation. Bauer (1943) compared line transects with cover estimates from quadrat measurements and concluded that in mixed-plant (grass, shrub, tree) communities, line sampling can be expected to give more accurate estimates of crown coverage than quadrat measurements.

Line Transect and Point Intercept. Point-intercept and line-transect methods are sometimes combined for estimating cover of plants in short vegetation types (Heady et al. 1959, Parker and Glendening 1942). Poissonet et al. (1972) used this method in tallgrass vegetation. A line transect is placed, and the intercept is read at randomly selected points. Bayonets, plumbs, long knives, and so forth have been used at sampling points to record the intercept. Gates et al. (1956) used a 30-m point transect, and a plumb was dropped from each 30-cm mark to record observations. In order to reduce bias, Tidmarsh and Havenga (1955) used a wheel without a rim, running on the points of the spokes. The wheel is rolled along a line, and observations are made where the spikes touch the ground. This method is not free of bias and cannot preclude the possibility of biased aiming by the observer. However, it compares well with the point-frame method for herbaceous plants.

The number and distance between points is an important consideration in sampling. It is usually more efficient to record more transects with fewer points per transect than fewer transects and more points per transect. However, the decision depends on the difference in time required to set up a transect compared to the time required to read additional points. Individual points may be located randomly or systematically. The individual transect is often used as the observation unit for statistical analysis. Calculations for line-point data are shown in Table 4.6. Refer to Table 4.4 for methods.

Line-point transects are useful for measuring changes in total vegetation cover. However, accuracy depends on the length of the line and number of points used per line. These transects are not suitable for measuring individual species cover where plants are intermingled and vegetation type boundaries are not distinct. Long lines produce underestimates of species cover when points are widely spaced, since several patterns in species are crossed. However, estimates of total cover are unaffected by length of line.

Spedding and Large (1957) used the point method to describe a sward in terms of height and cover. Each pin was graduated to measure all hits by species separately at each height on each pin. The number of hits at each height was converted into the number of hits per 100 points, which

TABLE 4.6   Example of Computations of Percent Cover with Line-Point Sampling Method Using a 100 m Transect and a Point Dropped at 10 cm Intervals

Species	Number of Hits per Line					Average Hits	Percent Cover	Percent Composition
	1	2	3	4	5			
Blue grama	219	382	71	106	318	219	21.9	41.5
Western wheatgrass	43	17	33	81	27	40	4.0	7.6
Needle-and-thread	64	73	0	42	9	38	3.8	7.2
Buffalograss	192	108	362	46	7	143	14.3	27.1
Sand dropseed	0	0	24	57	16	19	1.9	3.6
Sedge	76	62	12	119	42	62	6.2	11.7
Three awn	0	4	0	32	0	7	0.7	1.3
Total vegetation	594	646	502	483	419	528	52.8	100.0
Litter	187	132	272	231	318			
Bare ground	219	222	226	286	263			
Total	1000	1000	1000	1000	1000			

The data were recorded at Loamy Plains Range Site, near Fort Collins, Colorado (species same as in Table 4.4).

was plotted against height. This curve described the relationship between height and density for any given species. With this method, the mean height is

$$H = \frac{\Sigma dh}{t}$$
(4.23)

where t is the total number of hits at maximum cover, d is the difference between the number of hits at each centimeter and the number at the centimeter below it, and h is the height at which d occurred. The frame angle and the number of pins per frame can vary.

Fisser and Van Dyne (1966) examined the influence of the number and spacing of points on accuracy and precision of basal cover estimates based on point samples. Coefficients of variation were larger for random than for systematic spacing for all number of points (Table 4.7). Point sampling, as noted, usually requires more transects than line intercepts to obtain the same degree of accuracy.

Step Point. The step-point method of Evans and Love (1957) uses a single pin rather than pins grouped into a frame. An individual step point is established by the observer lowering the pin, guided by a notch on the toe of the observer's boot, to ground level. The boot is placed at an angle of approximately 30° to the ground to avoid disturbing plants around the point. The pin is lowered perpendicular to the sole of the boot, and the first hit by the pin or its sides is recorded. If no plant is hit, the pin is pushed into the ground, and the plant nearest to it in a forward direction (arc = 180°) is recorded. Evans and Love found that 300–500 points were needed to measure variability in cover for homogeneous vegetation. The

TABLE 4.7 Mean Values of Percent Frequency of Transect Occurrence, Percent Coefficient of Variation, and Number of Transects Required Averaged over all Plant Categories

	Point Samples					
	100 Points		50 Points		25 Points	
	S[a]	R[a]	S	R	S	R
Frequency transect occurrence	45	42	38	35	30	27
Coefficient of variation	201	212	223	238	256	283
Transects required[b]	120	133	150	165	204	249

From Fisser and Van Dyne. 1966. Influence of number and spacing of points on accuracy-precision. *J. Range Manage.* **19:** 209.

[a]S = systematic; R = random.

[b]Data are expressed as percent of the numbers required for the 500 unit lines.

step-point method compares favorably within 5% cover for grasses, but perhaps not for forbs. The step-point method, for example, requires about 30 min, while the point-frame method requires 3–4 hr to read cover from the same vegetation type. Most use of the step-point method is performed on annual or shortgrass grasslands because the method is limited for use in tall, heavy vegetation.

Subconscious selection of plants that affect any pin placement is a serious deficiency of the pins, not only in step point, but in most point-estimation methods. Owensby (1973) described a point-frame modification to eliminate subconscious bias in point placement and to make single point sampling easier. The basic design of the point frame is shown in Figure 4.10.

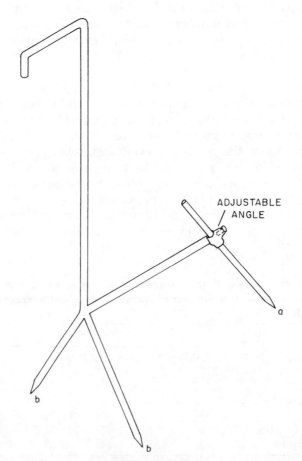

Figure 4.10 Point frame with a single point: (*a*) the point to be observed; (*b*) legs pressed into ground for stability (Modified and redrawn from Owensby. 1973. "Modified step point system for botanical composition and basal cover estimates." J. Range Manage. 26: 302.).

At an individual step point, one leg of a point frame is placed at the end of the observer's boot. The point frame is leaned backward toward the sampler on initial ground contact and is leaned forward until point contact is made with a plant crown or bare area. Hits are recorded the same way as in the step-point method. However, three persons are needed for efficient sampling and can read 3000 to 4000 points in a day.

4.2.5 Loop Method

It was pointed out previously that the thickness of the point is a major source of error in sampling cover or frequency by point methods. Parker (1951) found that enlarging the point contact area to as much as 1.9-cm-diameter circle reduced personal error compared to small points. He used this loop to measure condition and trend of herbaceous vegetation. However, loop frequency data are now used as indices of percent basal area or plant cover. The loop is systematically lowered along a permanent transect, and the presence or absence of species in the loop is recorded. The method is a compromise between a point quadrat, where only presence or absence of vegetation is recorded, and a larger quadrat where percent cover is estimated.

A number of studies have indicated that the loop method overestimates cover (Johnston 1957, Hutchings and Holmgren 1959, Parker and Harris 1959, Francis et al. 1972). The degree of overestimation is more for the single-stemmed plants or those with a small basal area. One must be aware of the limitations in use of loop frequency data in place of plant cover estimates. The relationship of frequency data to percent plant area depends on many attributes of vegetation other than the presence or absence of plants. Recall that frequency, as well as percent plant area, is affected by plant shape, size, and the distribution pattern.

Cook and Box (1961) found no significant differences between canopy cover estimates by loop and line-point methods on a mountain-brush-type vegetation in northern Utah. However, they obtained larger canopy cover values with the point-frame method compared to the loop method. The point-frame method resulted in a lower coefficient of variation compared to both the line-point and loop methods. On the contrary, Heady et al. (1959) observed satisfactory results in estimation of cover from loop sampling for dominant species in California. It appears that the loop method may give satisfactory results in some situations, but not in others. Therefore, it should be used only after proper comparison with a point method (Smith 1962).

4.2.6 Crown Diameter and Canopy Closure Method

Tree canopy cover or canopy closure can be estimated from measurements of crown diameter. In many species, crown diameter is correlated with the trunk diameter at breast height (dbh). This relationship makes it easier to estimate crown diameter and, in turn, canopy closure from measurement of dbh only. Thus, tree volume can also be estimated. In forest inventories, crown diameter is used as an independent variable in photostand volume tables. Dawkins (1963) collected data on 17 tropical tree species and examined the crown diameter:bole diameter relationships developed by other workers. He concluded that the most practical interpretation over the range of size for established trees was a straight line

$$y = a + bx \qquad (4.24)$$

where y is crown diameter, x is the dbh (diameter at breast height), a is a constant, and b is the regression coefficient. Where crown diameters are measured directly, measurements are usually taken in at least two directions (longest and shortest dimension) and then averaged.

Area of crown or foliar cover is estimated as

$$\text{Foliar cover} = \left(\frac{D_1 + D_2}{2} \right)^2 \pi \qquad (4.25)$$

where D_1 and D_2 are two measurements of crown diameter at right angles to each other. Because cover is expressed as a fraction or percent per unit area, unit area measurements are also taken simultaneously. Mueller-Dombois and Ellenberg (1974) consider this method impractical where a wide representative sample of cover by species is desirable over a large area. In these situations, use of the line-intercept method is more appropriate.

Crown diameters on aerial photographs can be measured with ordinary opaque rulers or transparent micrometer scales. Micrometer wedges are printed in fine lines on positive film and give accurate crown width measurements (Losee 1956). Then, photo measurements are converted to diameters in meters using the photograph scale ratio expressed in millimeters on the photographs.

Crown closure on an aerial photograph may be estimated visually by two methods (Pope 1960). The first is called "tree cramming." The scattered trees are ocularly moved to fill in openings among dense trees in the same way as that used for herbaceous cover. The proportion of an aerial photo occupied by trees constitutes crown closure. The second method is the "tree counting method." All trees of average diameter are counted. Then, the number of average diameter trees to fill in the open spaces are counted. The ratio of trees present to total count gives percent crown closure.

Johansson (1985) described a new approach to estimate canopy density by a vertical tubing method. The vertical tube consists of a 20-cm-long hand-held brass tube of 1-cm diameter. It is mounted on a universal joint so that it hangs vertically. There is a cross-hair at the upper end of the tube and a mirror at the lower end. The cross-hair is sighted through the mirror, and data are recorded for partial crown cover, full crown cover, or open sky. A "crown-free projection" (CFP) is calculated

$$\text{CFP} = \frac{S}{T} \times 100 \tag{4.26}$$

where S is number of open sky sightings and T is total number of observations.

Garrison (1949) described and used a device called the "moosehorn" to estimate crown closure. It is a type of periscope, and the top of the periscope has a grid of 25 dots on a glass plate. The periscope is held upright and fixed on a Jacob's staff. The instrument is completely leveled by a two-way level inside the periscope. The number of dots that are intercepted by a portion of the canopy are counted by viewing through a peephole. Note that 25 points provide an estimate of cover of ± 4 percent. If higher accuracy is needed, then more dots must be used.

4.2.7 Plotless Methods

Vegetation measurement methods involving collection of data from a plot area delimited in two dimensions (i.e., length and breadth) are grouped together as plot methods. However, some sampling techniques do not require a prescribed area unit and are referred to as plotless sampling methods. The point quadrat and the line transect are also, in a sense, plotless sampling units. There are two other important plotless techniques used for cover determination: Bitterlich's "variable-radius method" for tree basal area and shrub canopy area and the "point-centered quarter method" (PCQ) for determination of cover simultaneously with stem density.

4.2.8 Bitterlich's Variable-Radius Method

The variable plot method was developed by Bitterlich (1948) in Germany for determination of canopy coverage. Grosenbaugh (1952), Hyder and Sneva (1960), Cooper (1957, 1963), and Fisser (1961) extended application of this method for use in tree, shrub, and grassland types of vegetation. This method is called a "variable plot" because each plant counted represents a variable-sized plot.

Trees or shrubs are counted in a circle from a central sampling point with an angle gauge. Plants that are larger in diameter than the specified angle are included in the count while others are ignored. The number of plants counted are proportional to their stem or basal area per unit ground area. As explained in Chapter 2, the angle gauge is held horizontally with the cross-piece facing toward the plants. All plants surrounding the sample point are aimed at a fixed height, usually breast height (1.4 m) and viewed through the other end. Only those plants are counted for which diameter exceeds the cross-piece.

Bitterlich (1948) recommended a gauge ratio of 1.41 cm:100 cm, and the tree count is divided by 2 to obtain an estimate of basal area in square meters per hectare. Grosenbaugh recommended a still simpler ratio of 2 cm:100 cm, in which case the tree count is equal to basal area in square meters per hectare. This ratio has been used in several ecological studies (Shanks 1954, Rice and Kelting 1955, Rice and Penfound 1959).

4.2.9   Point-Centered Quarter Method (PCQ)

A number of distance methods have been developed for timber surveys to give density estimates (discussed in Chapter 5). However, in these methods, some additional data also can be collected to estimate basal area or canopy coverage by trees or foliage cover of shrubs. In the PCQ technique, four quarters are established at each sampling point. A cross of two lines, one in the direction of the compass and the other perpendicular to the compass line passing through the sample point, is established. The cross can also be randomly established by spinning a cross over each sampling point. Distance to midpoint of the nearest tree from the sample point and its diameter at breast height (dbh) is measured in each quarter. The density estimate, as obtained from equations in Chapter 5, is multiplied by the average basal area to give the basal area per unit area. The method can be used to measure cover of shrubs and trees. In the case of shrubs, instead of dbh for trees, two measurements of crown diameter are recorded at right angles from each other, as in the crown diameter method discussed earlier. Average foliar cover multiplied by density of shrubs gives total cover.

The PCQ method has some limitations in its application to the measurement of cover for shrubs and trees. If small shrubs or trees are obscured by larger plants, counting may be difficult. Some errors also result from plants with irregular outlines of canopies. It has been demonstrated that with only half as many observations, the same accuracy level was reached as with the line-intercept method. The method probably is not suitable if vegetation cover exceeds 35% because counting becomes difficult.

4.3   SEMIQUANTITATIVE METHODS

Methods discussed thus far involve measurements for determination of cover. These methods involve little or no judgment on the part of the investigator and are, therefore, regarded as quantitative methods. The results obtained are generally reproducible by other investigators. Many plant ecologists, however, are interested in qualitative attributes of species rather than quantitative attributes. Some qualitative methods can be used for mathematical computations of data and thus are suitable for statistical analysis. Therefore, such methods can be regarded as semiquantitative.

4.3.1 Cover Scale or Class Methods

The range of cover values, 0–100%, are arbitrarily divided into a number of categories, and each category is assigned a rating or scale value. In broadly defined cover classes, there is little chance for consistent human error in assigning coverage classes (Daubenmire 1959). Most scales have unequal class intervals, which allow easier estimation of species cover to area relationship. A finer breakdown of scale toward the lower scale values allows better estimation of less abundant species. The species occurring in the sample plot are assigned these ratings or scale values on the basis of the area occupied by species. Some of these methods also take into account the numbers of each species, in addition to the area occupied by it, to obtain the "species magnitude" (Braun-Blanquet 1965) or "species significance" (Krajina 1960). The midpoints of each class can be used for statistical analysis of data. The use of midpoints for interpretation is based on the assumption that actual values tend to be symmetrically dispersed about these points. This assumption is a strong one and, if violated, renders any statistical analysis useless.

In order to assign scale or rating values to each species, the observer makes a guess whether a species occupies more or less than 50% of the area of the sample plot. If it occupies less than 50%, the next question to ask is whether it occupies more than or less than 25%. Similarly, if it occupies more than 50%, it is determined whether it occupies more than or less than 75% of area. Similar reasoning is continued until the range of the species cover value is narrowed down to a predetermined rating or scale.

Even an experienced investigator may assign a species to a lower or a higher scale value than occupied by the species. Since the midpoints of each class interval are wide apart, there can be a large variation in data between investigators. Also, in a species-rich herbaceous community, errors of estimate are more likely with finer-scale intervals than with scales that have broad categories.

Braun-Blanquet Scale. The Braun-Blanquet scale, a semiquantitative method, gives a combined estimate of abundance and cover. The sampling unit is called a "releve" and its size is based on the minimal area concept. The first step is to familiarize oneself with vegetation of the releve. The

TABLE 4.8 The Braun-Blanquet Cover-Abundance Scale

Rating	Number of Plants	Area Occupied by a Species
+	Sparsely or very sparsely present	Very small
1	Plentiful	Small
2	Very numerous	10–25%
3	Any number	25–50%
4	Any number	50–75%
5	Any number	>75%

From J. Braun-Blanquet. 1964. *Pflanzensoziologie*. Reprinted by permission of Springer-Verlag, Inc., New York.

second step is to prepare a species list. The third step is to assign a value to each species from the Braun-Blanquet cover-abundance scale (Table 4.8). A species with less than 5% but more than 1% cover receives the value 1, and those with less than 1% receive a + (plus). The rare species receive a notation of *r*.

Daubenmire Scale. Daubenmire (1959) proposed a scale (Table 4.9) for estimating cover alone. A 0.1-m^2 plot (20 × 50-cm inside dimensions) was found to be satisfactory. The method is also referred to as the "canopy coverage method." The frame (5-mm steel) is painted so as to divide it into quarters, crosswise. In one corner of the frame, two sides of a 71-mm^2 area are painted (Fig. 4.11). This allows reference areas equal to 5, 25, 50, 75, and 95% of the frame. Sharpened legs, 3 cm long, may be welded to each corner for holding the frame in place. The legs, acting as points, also permit tallying the condition of the ground surface (bare ground, litter, etc.), which adds to this technique the chief advantage of the point-frequency method (Daubenmire 1959). Bailey and Poulton (1968) modified the Daubenmire

TABLE 4.9 The Daubenmire Cover Scale

Cover Class	Range of Cover (%)	Class Midpoints (%)
1	0–5	2.5
2	5–25	15.0
3	25–50	37.5
4	50–75	62.5
5	75–95	85.0
6	95–100	97.5

From Daubenmire. 1959. A canopy coverage method. *N. W. Sci.* **33**: 43–64.

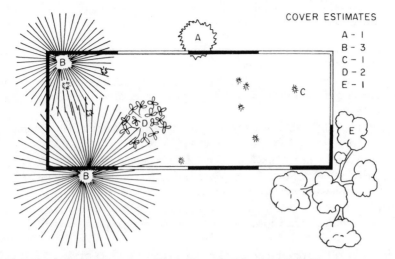

Figure 4.11 Daubenmire cover frame (From R. F. Daubenmire. "A canopy coverage method." Northwest Science. 33: 52.).

cover scale by separating 0–5% cover class into two classes, one for a range of 0–1% (midpoint 0.5%) and the second for 1–5% (midpoint 3.0%).

Domin–Krajina Cover-Abundance Scale. The Domin–Krajina cover-abundance scale (Table 4.10) was originally developed by Domin and later modified by Krajina (1933). The method is useful in forest communities where differences in abundance among rarer species are often quite noticeable. The method was successfully applied in many studies of forest communities in British Columbia (Krajina 1969) and also by Kershaw (1968) in Nigeria.

TABLE 4.10 The Domin–Krajina Cover-Abundance Scale

Rating	Number of Plants of Each Species	Cover (%) of the Species
10	Any number	100
9	Any number	>75, but <100
8	Any number	50–75
7	Any number	33–50
6	Any number	25–33
5	Any number	10–25
4	Any number	5–10
3	Scattered	1–5
2	Very scattered	<1
1	Seldom	Insignificant
+	Solitary	Insignificant

4.4 BIBLIOGRAPHY

Aberdeen J. E. C. 1957. The uses and limitations of frequency estimates in plant ecology. *Aust. J. Bot.* **5**: 86–102.

Bailey, A. W., and C. E. Poulton. 1968. Plant communities and environmental relationships in a portion of the Tillamook burn, northwestern Oregon. *Ecology* **49**: 1–13.

Bauer, H. L. 1943. The statistical analysis of chaparral and other plant communities by means of transect samples. *Ecology* **24**: 45–60.

Bitterlich, W. 1948. Die Winkelzahlprobe. *Allg. Forst-und-Holzwirtsch. Ztg.* **59**: 4–5.

Blackman, G. E. 1935. A study of statistical methods of the distribution of species in grassland associations. *Ann. Bot.* **49**: 749–777.

Bonner, G. M. 1964. The influence of stand density on the correlation of stem diameter with crown width and height for lodgepole pine. *Forestry Chron.* **40**: 347–349.

Booth, W. E. 1943. Tripod method of making chart quadrats. *Ecology* **24**: 262.

Borman, F. H., and M. F. Buell. 1964. Old-age stand of hemlock–northern hardwood forest in central Vermont. *Bull. Torrey Bot. Club* **91**: 451–465.

Braun-Blanquet, J. 1965. *Plant Sociology: The Study of Plant Communities.* (Translated, revised and edited by C. D. Fuller and H. S. Conard.) Hafner, London, 439 pp.

Buell, M. F. and J. E. Cantlon. 1950. A study of two communities of New Jersey Pine Barrens and a comparison of methods. *Ecology* **31**: 567–586.

Burzlaff, D. F. 1966. The focal-point technique of vegetation inventory. *J. Range Manage.* **19**: 222–223.

Cain, S. A., and G. M. de Oliveira Castro. 1959. *Manual of Vegetation Analysis.* Harper, New York, 325 pp.

Canfield, R. H. 1941. Application of the line interception method in sampling range vegetation. *J. Forestry* **39**: 388–394.

Chan, S. S., R. W. McCreight, J. D. Walstad, and T. A. Spies. 1986. Evaluating forest vegetative cover with computerized analysis of fisheye photographs. *Forestry Sci.* **32**: 1085–1091.

Cook, C. W., and T. W. Box. 1961. A comparison of the loop and point methods of analyzing vegetation. *J. Range Manage.* **14**: 22–27.

Cooper, C. F. 1957. The variable plot method for estimating shrub density. *J. Range Manage.* **10**: 111–115.

————. 1963. An evaluation of variable plot sampling in shrub and herbaceous vegetation. *Ecology* **44**: 565–569.

Crocker, R. L., and N. S. Tiver. 1948. Survey methods in grassland ecology. *J. Br. Grassl. Soc.* **3**: 1–26.

Culley, M. 1938. Densimeter, an instrument for measuring the density of ground cover. *Ecology* **19**: 588–590.

Curtis, J. T., and R. P. McIntosh. 1950. The interrelations of certain analytic and synthetic phytosociological characters. *Ecology* **31**: 434–455.

Daubenmire, R. F. 1959. Canopy coverage method of vegetation analysis. *Northwest Sci.* **33**: 43–64.

————. 1968. *Plant Communities: A Textbook of Plant Synecology.* Harper & Row, New York, 300 pp.

Dawkins, H. C. 1963. Crown diameters: Their relation to bole diameter in tropical forest trees. *Commonwealth Forestry Rev.* **42**: 318–333.

Dilworth, J. R., and J. F. Bell. 1973. *Variable Probability Sampling—Variable Plot and Three-P. A Pocket Book.* Oregon State University Bookstore, Corvallis, OR, 129 pp.

Evans, R. A., and R. M. Love. 1957. The step-point method of sampling: A practical tool in range research. *J. Range Manage.* **10**: 208–212.

Fisser, H. G. 1961. Variable plot, square foot plot, and visual estimate for shrub crown cover measurements. *J. Range Manage.* **14**: 202–207.

Fisser, H. G., and G. M. Van Dyne. 1966. Influence of number and spacing of points on accuracy and precision of basal cover estimates. *J. Range Manage.* **19**: 205–211.

Francis, R. E., R. S. Driscoll, and J. N. Reppert. 1972. *Loop-Frequency as Related to Plant Cover, Herbage Production, and Plant Density.* USDA Forest Service Research Paper RM-94.

Garrison, G. A. 1949. Uses and modifications for the "moosehorn" crown closure estimator. *J. Forestry* **47**: 733–735.

Gates, D. H., L. A. Stoddart, and C. W. Cook. 1956. Soil as a factor influencing plant distribution on salt-deserts of Utah. *Ecol. Monogr.* **26**: 155–175.

Goodall, D. W. 1952. Some considerations in the use of point quadrats for the analysis of vegetation. *Aust. J. Sci. Res., Ser. B* **5**: 1–41.

————. 1953. Point–quadrat methods for the analysis of vegetation. *Aust. J. Bot.* **1**: 457–461.

Greig-Smith, P. 1957. *Quantitative Plant Ecology.* Academic Press, New York, 198 pp.

————. 1983. *Quantitative Plant Ecology,* 3rd ed. Univ. of California Press, Berkeley, 359 pp.

Grosenbaugh, L. R. 1952. Plotless timber estimates—new, fast, easy. *J. Forestry* **50**: 32–37.

Hasel, A. A. 1941. Estimation of vegetation-type area by linear measurement. *J. Forestry* **39**: 34–40.

Heady, H. F., and L. Rader. 1958. Modifications of the point frame. *J. Range Manage.* **11**: 95–96.

Heady, H. F., R. P. Gibbens, and R. W. Powell. 1959. A comparison of the charting, line intercept, and line point methods of sampling shrub types of vegetation. *J. Range Mange.* **12**: 180–188.

Hector, J. M., and L. F. Irvine. 1938. A new apparatus for charting vegetation. *S. Afr. J. Sci.* **35**: 231–235.

Hill, A. A. 1920. Charting quadrats with a pantograph. *Ecology* **1**: 270–273.

Holscher, C. E. 1959. General review of methodology on use of plant cover and composition for describing forest and range vegetation, pp. 39–44. In *Techniques and Methods of Measuring Understory Vegetation.* USDA Forest Service Southern Forestry Experiment Station and Southeastern Forestry Experiment Station.

Hormay, A. L. 1949. Getting better records of vegetative changes with the line interception method. *J. Range Manage.* **2**: 67–69.

Husch, B., C. I. Miller, and T. W. Beers. 1982. *Forest Mensuration.* Wiley, New York, 402 pp.

Hutchings, S. S., and R. C. Holmgren. 1959. Interpretation of loop-frequency data as a measure of plant cover. *Ecology* **40**: 668–677.

Hyder, D. N., and F. A. Sneva. 1960. Bitterlich's plotless method for sampling basal ground cover of bunch grasses. *J. Range Manage.* **13**: 6–9.

Hyder, D. N., R. E. Bement, E. E. Remmenga, and C. Terwilliger, Jr. 1965. Frequency sampling of blue grama range. *J. Range Manage.* **18**: 90–94.

Ibrahim, K. M. 1971. Ocular point quadrat method. *J. Range Manage.* **24**: 312.

Jackson, M. T., and R. O. Petty. 1973. A simple device for measuring vertical projection of three crowns. *Forestry Sci.* **19**: 60–62.

Johansson, T. 1985. Estimating canopy density by the vertical tube method. *Forest Ecol. Manage.* **11**: 139–144.

Johnson, L. 1927. An instrument for list charting. *Ecology* **8**: 282–283.

Johnston, A. 1957. A comparison of the line interception, vertical point quadrat, and loop methods as used in measuring basal area of grassland vegetation. *Can. J. Plant Sci.* **37**: 34–42.

Kemp, C. D., and A. W. Kemp. 1956. The analysis of point quadrat data. *Aust. J. Bot.* **4**: 167–174.

Kershaw, K. A. 1968. A survey of the vegetation in Zaria Provice, N. Nigeria. *Vegetatio* **15**: 244–268.

Kinsinger, F. E., R. E. Eckert, and P. O. Currie. 1960. A comparison of the line interception, variable-plot, and loop methods as used to measure shrub-crown cover. *J. Range Manage.* **13**: 17–21.

Krajina, V. J. 1933. Die Pflanzengesellschaften des Mlyncia-Tales in den Vysoke Tatry (Hohe Tatra). Mit besonderer Beruckscichtigung der Okologischen Verhaltnisse. *Botan. Centralbl., Beih., Abt. II* **50**: 774–957; **51**: 1–224.

———. 1960. Can we find a common platform for the different schools of forest type classification? *Silva Fennica* (Helsinki) **105**: 50–59.

———. 1969. Ecology of forest trees in British Columbia. *Ecology of Western North America* (published by the Department of Botany, University of British Columbia), Vol. 2, pp. 1–147. Including map of "Biogeoclimatic Zones."

Levy, E. B., and E. A. Madden. 1933. The point method for pasture analysis. *N. Z. J. Agric.* **46**: 267–279.

Lindsey, A. A. 1955. Testing the line-strip method against full tallies in diverse forest types. *Ecology* **36**: 485–495.

Losee, S. T. B. 1956. Measurement of stand density in forest photogrammetry. Canadian Institute of Surveying and Programming. (Paper presented.)

Morrison, R. G., and G. A. Yarranton. 1970. An instrument for rapid and precise point sampling of vegetation. *Can. J. Bot.* **48**: 293–297.

Mueller-Dombois, D., and H. Ellenberg. 1974. *Aims and Methods of Vegetation Ecology.* Wiley, New York, 547 pp.

Murray, J. M. 1946. *A Guide to Botanical Techniques for Pasture Research Workers in South Africa*. South African Department of Agriculture, 30 pp.

Nerney, N. J. 1960. A modification for the point-frame method of sampling range vegetation. *J. Range Manage.* **13**: 261–262.

Oosting, H. J. 1956. *The Study of Plant Communities: An Introduction to Plant Ecology*, 2nd ed. Freeman, San Francisco, 440 pp.

Owensby, C. E. 1973. Modified step-point system for botanical composition and basal cover estimates. *J. Range Manage.* **26**: 302–303.

Parker, K. W. 1951. *A Method of Measuring Trend and Range Condition on National Forest Ranges*. USDA Forest Service, Washington, DC, 26 pp.

Parker, K. W., and G. E. Glendening. 1942. *General Guide to Satisfactory Utilization of the Principal Southwestern Range Grasses*. Research Note 104, U.S. Department of Agriculture, Forest Service, Southwestern Forest Range Experiment Station, p. 4.

Parker, K. W., and R. W. Harris. 1959. *The 3-Step Method for Measuring Condition and Trend of Forest Ranges: A Resume of Its History, Development, and Use*. U.S. Forest Service South and Southeast *Forestry Experiment Station Proceedings*, pp. 55–69.

Parker, K. W., and D. A. Savage, 1944. Reliability of the line interception method in measuring vegetation on the Southern Great Plains. *J. Am. Soc. Agron.* **36**: 97–110.

Pearse, K. 1935. An area-list method of measuring range plant populations. *Ecology* **16**: 573–579.

Pearse, K., J. F. Pechanec, and G. D. Pickford. 1935. An improved pantograph for mapping vegetation. *Ecology* **16**: 529–530.

Pierce, W. R., and L. E. Eddleman. 1970. A field stereo photographic technique for range vegetation analysis. *J. Range Manage.* **23**: 218–220.

Pierce, W. R., and L. E. Eddleman. 1973. A test of stereo-photographic sampling in grasslands. *J. Range. Manage.* **26**: 148–150.

Poissonet, P. S., P. M. Daget, J. A. Poissonet, and G. A. Long. 1972. Rapid point survey by bayonet blade. *J. Range Manage.* **25**: 313.

Pope, R. B. 1960. Ocular estimation of crown density on aerial photos. *Forestry Chron.* **36**(1): 89–90.

Ratcliff, R. D., and S. E. Westfall. 1973. A simple stereo-photographic technique for analyzing small plots. *J. Range Manage.* **26**: 147–148.

Raunkiaer, C. 1934. *The Life Forms of Plants and Statistical Plant Geography*. Oxford University Press, Oxford, 632 pp.

Rice, E. L., and R. W. Kelting. 1955. The species area curve. *Ecology* **36**: 7–11.

Rice, E. L., and W. T. Penfound. 1959. The upland forest of Oklahoma. *Ecology* **40**: 593–608.

Romell, L. G. 1930. Comments on Raunkiaer's and similar methods of vegetation analysis and the "law of frequency." *Ecology* **11**: 589–596.

Savage, D. A., and L. A. Jacobson. 1935. The killing affect of heat and drought on buffalograss and blue grama grass at Hays, Kansas. *J. Am. Soc. Agron.* **27**: 566–582.

Shanks, R. E. 1954. Plotless sampling trials in Appalachian forest types. *Ecology* **35:** 237–244.

Smith, J. G. 1962. An appraisal of loop transect method for estimating root crown area changes. *J. Range Manage.* **15:** 72–78.

Smith, J. H. G. 1983. Recent analysis of crown cover define opportunities for increasing stocking of British Columbia Forests. *Forestry Chron.* **59:** 262.

Society of American Foresters. 1958. *Forestry terminology—a Glossary of Technical Terms Used in Forestry.* Society of American Foresters, Washington, DC, 97 pp.

Spedding, C. R. W. and R. W. Large. 1957. A point-quadrat method for the description of pasture in terms of height and density. *J. British Grassland Soc.* **12:** 229–234.

Stanton, F. W. 1960. Ocular point frame. *J. Range Manage.* **13:** 153.

Stephenson, S. N., and M. F. Buell. 1965. The reproducibility of shrub cover sampling. *Ecology* **46:** 379–380.

Tansley, A. G., and T. F. Chipp (eds.). 1926. *Aims and Methods in Study of Vegetation.* British Empire Veg Chee and Crown Agents for the Colonies, London, 383 pp.

Tidmarsh, C. E. M., and C. M. Havenga. 1955. *The Wheel-Point Method of Survey and Measurement of Semi-open Grasslands and Karoo Vegetation in South Africa.* Botanical Survey South Africa, Memoir 29, 49 pp.

Tinney, F. W., O. S. Aamodt, and H. L. Ahlgren. 1937. Preliminary report of a study on methods used in botanical analyses of pasture swards. *J. Am. Soc. Agron.* **29:** 835–840.

Warren-Wilson, J. 1959a. Analysis of spatial distribution of foliage by two-dimensional point quadrats. *New Phytol.* **58:** 92–101.

———. 1959b. Analysis of the distribution of foliage area in grassland, pp. 51–61. *The Measurement of Grassland Productivity,* J. D. Ivins (ed.). Butterworths, London.

———. 1960. Inclined point quadrats. *New Phytol.* **59:** 1–8.

———. 1963a. Estimation of foliage denseness and foliage angle by inclined point quadrats. *Aust. J. Bot.* **11:** 95–105.

———. 1963b. Errors resulting from thickness of point quadrats. *Aust. J. Bot.* **11:** 178–188.

Weaver, J. E. and F. W. Albertson. 1943. Resurvey of grasses, forbs, and underground plant parts at the end of the great drought. *Ecol. Monogr.* **13:** 63–117.

Wein, R. W., and A. N. Rencz. 1976. Plant cover and standing crop sampling procedures for the Canadian High Arctic. *Arctic Alpine Res.* **8:** 139–150.

Wells, K. F. 1971. Measuring vegetation changes on fixed quadrats by vertical ground stereo-photography. *J. Range Manage.* **24:** 233–236.

Wenger, K. F. 1956. Growth of hardwoods after clear cutting loblolly pine. *Ecology* **37:** 735–742.

Whitman, W. C., and E. I. Siggeirsson. 1954. Comparison of line interception and point contact methods in the analysis of mixed grass range vegetation. *Ecology* **35:** 431–436.

Whittaker, R. H. 1970. *Communities and Ecosystems.* MacMillan, Collier-MacMillan Ltd., London, 162 pp.

Wimbush, D. J., M. D. Barrow, and A. B. Costin. 1967. Color stereo-photography for the measurement of vegetation. *Ecology* **48:** 150–152.

Winkworth, R. E. 1955. The use of point quadrats for the analysis of heathland. *Aust. J. Bot.* **3:** 68–81.

Winkworth, R. E., and D. W. Goodall. 1962. A crosswire sighting tube for point quadrat analysis. *Ecology* **43:** 342–343.

Winkworth, R. E., R. A. Perry, and C. O. Rossetti. 1962. A comparison of methods of estimating plant cover in an arid grassland community. *J. Range Manage.* **15:** 194–196.

Wright, R. G. 1972. Computer processing of chart quadrat maps and their use in plant demographic studies. *J. Range Manage.* **25:** 476–478.

5 DENSITY

"Density" is defined as the number of individuals in a given unit of area or the reciprocal of the mean area of space per individual. Density is generally considered to be a readily obtainable and easily understood characteristic of vegetation. The number of individuals is one of the most useful animal population attributes because individuals are easily identified, and members of the same species are of similar size (Pieper 1973). Difficulties often arise, however, when density is used to describe vegetation communities. The purpose of this chapter is to discuss methods used to estimate plant density.

Density is frequently used to describe the vegetation characteristics of a community. Yet, there are several problems encountered in obtaining an accurate estimate of density; these include definition of an individual plant, size and shape of a quadrat with associated boundary error, and use of estimates from plotless methods.

5.1 RELATED MEASUREMENTS

5.1.1 Cover

Confusion exists in some literature between the terms density and cover (Cook 1962, Pieper 1973). An early definition of density was given by Dayton (1931) as the relative degree to which vegetation covers the ground surface. Dayton's definition of density, however, is used by most plant ecologists when they refer to plant cover. Carpenter (1938) first defined density, as it is used today, as the exact ratio between the number of individuals of the same species observed on a certain surface and the extent of that surface. The Committee on Nomenclature of the Ecological Society of America (1952) defined density as "the relation between the number and/or volume of individuals of a species (or all species) on an area; or more correctly, in a space, refers to the closeness of individuals to another." Thus, density is defined as either the total number of individuals in a specific area or the reciprocal of the mean area or space per individual.

5.1.2 Frequency

Another term closely related to density is "frequency." The concept of frequency, first developed and used by the Danish ecologist C. Raunkiaer, refers to the number of times one may encounter a species when walking through a stand (Stearns 1959). Frequency is defined as the relation between the number of sampling units in which the species is present and the total number of sampling units. This is often expressed as a percentage. That is, frequency is the percentage of sample units that contain the species. As the size of the plot is increased, the chance of a species occurring within that plot also is increased. Then, frequency is actually dependent on density and pattern of a species and, thus, has meaning only in relation to the particular size and shape of the area of the plot used (Greig-Smith 1964).

5.1.3   Abundance

The concept of abundance of species has both a qualitative and quantitative meaning. "Abundance" refers to an arbitrarily estimated range in numerical values which expresses plentifulness or scarcity of a species. These ranges are usually expressed by assigning the species to abundance classes. Braun-Blanquet (1932) listed five such abundance classes: (1) very sparse, (2) sparse, (3) not numerous, (4) numerous, and (5) very numerous.

These classes are relative in meaning and may have little consistency from one investigator to another, or from one study area to another (Hope-Simpson 1940, Smith, 1944). The measurement of abundance, then, is subject to personal bias. Some plant species, because of their form, color, economic importance, aesthetic appeal, or familiarity to the observer, may be overestimated. Reversal of some or all of these characteristics may also lead to an underestimate of how plentiful a species really is.

On the basis of the foregoing, abundance should have a quantitative meaning. An appropriate definition, then, is the average number of individuals of a species per unit area for plots that contain the species. The definition is illustrated in Figure 5.1. In the example, abundance is 4.0 m^{-2}, since only three quadrats contain the 12 individiuals. Density is 2.4 individuals/m^2 (i.e., 12/5). Then, abundance also includes a combination of both frequency and density estimates.

5.2   LIMITATIONS OF THE DENSITY ESTIMATE

One must be able to recognize and define individual plants since this is critical to the measurement of plant density. Yet, identification of an individual plant is often one of the greatest difficulties in the determination of plant density. For instance, an individual plant may be defined as the aerial parts corresponding to a single root system (Strickler and Stearns 1962). This is easy for trees and other single-stemmed species, such as annuals, as they are readily determined. This may not always be the case, however, since annual and perennial grasses, when excavated, may show

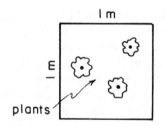

Figure 5.1   Abundance is equal to 12 divided by 3 ($A = 4.0$ plants/m^2).

a wide range in variability in the number of culms for each root system. On the other hand, bunchgrasses may form easily recognized individuals.

Studies of plants throughout the world have noted, however, that old and large bunchgrasses frequently break into smaller individual groups, thus making the determination of individual plants difficult (e.g., Walker 1970). Moreover, perennial grasses and forbs sometimes spread vegetatively (i.e., stoloniferous or rhizomatous forms) and are impossible to recognize as individuals. Shrubs that are close together and produce both single- and multiple-stemmed plants also make density estimation difficult. If the definition of the individual is difficult, then some method may be used which sets arbitrary limits to either the rooting areas or the aboveground parts of an individual. Brown (1954) noted that these arbitrary limits vary in size according to life-forms, and any modification made in the unit definition will prevent comparison with density estimates obtained by other units used for counts.

Again, it must be emphasized that the only practical definition of the counting unit for plants is the individual, whether it be a stalk, a culm, or a bunch of culms. The value of defining individuals depends on the purpose of the study, definition of the unit, and precision of the count. Use of a stem or shoot, as the individual, has the advantage of less variation in unit size than is the case with the use of either bunch or rhizomatous plants. Furthermore, the number of stems or culms may have a higher correlation with other measurements, such as biomass amount or basal cover.

It should be apparent by now that a knowledge of the morphology and ecology of a species is necessary before an estimate of density can be made. Examination of the root system of the species may be necessary to determine what constitutes an individual. This may be true if a plant unit is difficult to identify or if the vegetation contains mixed life-forms such as perennial grasses and forbs. Then, other vegetation characteristics, such as cover or biomass amount, may be more easily measured.

A density estimate alone is a limited measure to determine plant community dominance. That is, density estimates only the number of individuals per unit area and not the relative contribution of each species to total plant biomass produced in the vegetation type. Species contribution to total biomass may vary significantly from year to year as a result of differing environmental conditions. Yet, density may not change. In contrast, annual plants will vary in number of individuals as annual precipitation and temperature vary. On the other hand, numbers of perennial plant species tend to remain fairly constant over time. This is true for shrubs, trees, and some perennial grasses and forbs.

In spite of density estimate limitations, counting remains one of the easiest quantitative species measure to understand. Estimates of density are useful for monitoring plant responses to various vegetation treatments such as defoliation or environmental perturbations.

Four major techniques are used to estimate density: plot, distance, line transect, and indirect methods based on frequency observations. The plot and distance techniques are often most used to obtain estimates of species densities.

The following sections represent a fairly complete presentation of density estimation techniques. Some discussion will include a limited use of theoretical ideas regarding pattern analysis, distance distributions, and species–area relations. Therefore, this chapter is somewhat more difficult, in places, for the reader who has not acquired a strong statistical background. Yet, inclusion of this material is necessary for certain techniques so that the reader can develop a better understanding of the technique and its provision of a density estimate.

5.3   QUADRAT TECHNIQUES

The number of individuals found within plots may be counted to estimate density. Pound and Clements (1898) were among the first to use quadrats to measure vegetation characteristics. They used 5-m² frames to count plant species in some vegetation types in Nebraska. Pound and Clements (1898) also realized that a problem in the estimation of density is the degree of dispersion of individuals and quadrat size and shape. Thus, if quadrat techniques are used to obtain density estimates, three characteristics must be considered before sampling is initiated: (1) distribution of the plants, (2) size and shape of plot, and (3) the number of observations needed to obtain an adequate estimate of density.

5.3.1   Distribution

Plants that are randomly distributed follow the Poisson series. That is a rare-events mathematical series. If a species is rare in its occurrence in a quadrat, then the accuracy of the density estimate depends only on the number of individuals counted. This relationship is more fully appreciated by noting the following equations (Greig-Smith 1964). Let x be the number of individuals that are counted in n quadrats. Then,

$$\text{Density} = \frac{x}{n} \tag{5.1}$$

The variance s^2 is equal to its mean in a Poisson distribution. Then

$$s^2 = \frac{x}{n} \tag{5.2}$$

and the variance of the mean of n quadrats is

$$s_{\bar{x}}^2 = \frac{1}{n}\left(\frac{x}{n}\right) = \frac{x}{n^2} \tag{5.3}$$

The standard error of the mean is

$$s_{\bar{x}} = \frac{\sqrt{x}}{n} \tag{5.4}$$

The ratio of standard error of mean to the mean then is

$$\left(\frac{\sqrt{x}}{n}\right)\left(\frac{n}{x}\right) = \frac{1}{\sqrt{x}} \tag{5.5}$$

Then it is realized that the standard error of the density estimate remains the same for the same number of individuals counted. This is true regardless of the size or number of quadrats used. Since a single observation from a randomly distributed population can be regarded as an unbiased estimate of the population parameter, the sample variance and standard error can be estimated from the number of individuals that occurred in the single quadrat. Let x be the number of individuals in a single quadrat as before. Then, x and $\sqrt{x}$ are estimates of the variance and the standard error, respectively. Note, however, that this single observation may deviate significantly from the true mean and, therefore should be regarded as only a rough approximation of the population mean. In most natural populations, individual plants are not randomly located, but occur in clumps. In fact, random populations of plants are the exceptions in nature, but may be found in single-species populations occurring in pure stands (Goodall 1952, Hutchings and Morris 1959, Strickler and Stearns 1962).

In populations of nonrandomly distributed individuals, density is difficult to evaluate by simple techniques since the data distribution may not follow a Poisson series. One should be aware that quadrat size affects the distribution (whether Poisson, binomial, etc.) of the data. For example, a large quadrat size will result in a more normal distribution of the data and reduce the magnitude of the variance in comparison to a small quadrat. Intuitively, it is seen that more individual plants may occur in a large quadrat compared to a small quadrat, which allows for the occurrence of few individuals. The latter provides a "rare-events" form for a distribution, that is, a Poisson distribution.

5.3.2 Quadrat Size and Shape

Determination of the proper quadrat size for a plant species population is difficult. One should consider several factors before selecting a given plot size and shape. Some factors were presented in Chapter 2. First, recall that a plot has a greater length of boundary per unit area than successive larger ones. The smaller the plot, the greater the opportunity for boundary error. That is, an observer may consistently include individual plants near or on the boundary that should not have been counted, or may not count those that were within, but near the boundary. For this reason, a quadrat should not be small if density estimates are made. Second, the shape of the distribution curve (see Chapter 3) affects significance tests which are based on a normal distribution. Plot size does affect the curve shape, and an adequate plot size will decrease the variance for numbers of plants between observations. Data on density will then approximate a normal distribution curve. Third, field experience will indicate, even to the novice, that counting large numbers of individual plants can be tedious and lead to inaccurate counts in large plots. A large quadrat can be subdivided or individuals marked as they are counted, but the purpose of using a large quadrat is then defeated. Subdivision of such quadrats only leads to increasing boundary lengths per unit area of new subquadrats. This problem can be solved by increasing the number of quadrats used for counting.

Bartlett (1948) found that the most efficient size of quadrats for estimation of density corresponds to about 20% absence. This value for absence is the same as the product of quadrat size and plant density of about 1.6. This constant, 1.6, is obtained from the relationship between density and frequency of a species found in a random distribution. If

$$\text{Density} = -\ln\left(1 - \frac{F}{100}\right) \qquad (5.6)$$

where F is the percentage frequency, and ln is the natural logarithm, then $1.0 - 0.80 = 0.20$, which is the absence of the species. Then,

$$\text{Density} = -\ln(1.0 - 0.80) = -\ln(0.20) = 1.61$$

Bartlett suggested that the efficiency is maintained over a size range of area, 0.7–3.0 m². This method of determining quadrat size is applicable when the distribution of plants is random or nearly so and may be used when the relationship between the logarithm of percent absence and plant density is linear.

Quadrat size and shape is known to affect the accuracy of a sample estimate of density. Van Dyne et al. (1963) summarized studies on quadrat sizes and shapes on western U.S. grassland vegetation. Boundary length:area ratios were noted to be lowest in circular quadrats, and this ratio decreases as the quadrat size increases. Many ecologists have observed that more species are included in long, narrow quadrats because of the tendency for vegetation to be clumped. In fact, optimum quadrat size and shape depends on the distribution of the species measured, and in general, larger quadrats are recommended for use in sparse vegetation. Although small plots are generally more efficient statistically, they often yield skewed data for density. The greatest sampling efficiency is generally indicated by the plot that provides the smallest variance for any specific measurement. Bormann (1953) noted, for rectangular quadrats, that the variance is reduced only if the long axis of the quadrat crosses any banding in vegetation pattern. Such a perpendicular alignment of a quadrat increases the variation within the sample unit but decreases the variance between units. That is, extremes in counts of plants will cease to occur and, instead, a smaller range in numbers will exist. Therefore, the unit contains more plants, but the variance between observations is decreased.

Eddleman et al. (1964) tested plot sizes and shapes to estimate density and frequency in dry-mesic alpine vegetation in northern Colorado. Plots were 100, 400, 800, and 1600 cm^2 with rectangular, square, and circular shapes. Two estimates of efficiency were made. One used the time required to sample for a 10% standard error of the mean. The other was the product of the average weighted standard deviation for a particular plot shape and size multiplied by the time required to read sufficient plots within 10% of the mean density (Table 5.1).

Plots that required the fewest number were a 400-cm^2 circle, 40 × 40 cm^2, and 10 × 40-cm rectangle. Rectangular quadrats of the same area as square plots needed fewer observations in two out of three cases. Rectangular quadrats were more efficient for some species, while square ones were more efficient for others. Standard deviations increased with increase in quadrat size, but when adjusted for quadrat size, larger quadrats had smaller standard deviations. Time to sample was less for medium to large plots and for rectangular plots. The 400 cm^2 plots of the Eddleman et al. study are favored because there is less likelihood of miscounting species.

Evans (1952) discussed the influence of quadrat size on the distributional patterns of plant populations. Quadrats of sizes: 16, 8, 4, 2, 1, $\frac{1}{2}$, $\frac{1}{4}$, $\frac{1}{8}$, and $\frac{1}{16}$ m^2 were used. The species studied had aggregated populations, and density estimates on an area basis were found to be the same for all quadrat sizes but increased proportionately for larger quadrats when compared on a quadrat basis.

Frequency of species can be compared with a Poisson series to detect spatial patterns, and Evans (1952) found that almost all species and quadrat sizes showed departure from randomness. One exception was the smallest-sized quadrat ($\frac{1}{16}$ m^2), where the occurrence of more than one individual per quadrat was extremely rare. The $\frac{1}{16}$-cm^2 quadrats probably approached the minimum area needed. That is the mean area per plant. This size of plot also caused spatial exclusion to occur. This happens when only one

TABLE 5.1   Efficiency Estimates and Ranking of Quadrats

Plot[a]	Average Number of Plots for Density Stabilitiy Points	Time (min) to Read Plots for Density Stability Points	Average Weighted Standard Deviation
10 × 40 cm R	51	69	5.08
20 × 80 cm R	56	120	3.44
40 × 40 cm R	48	100	3.47
400 cm^2 C, R	40	54	5.93
20 × 20 R	63	85	6.68
20 × 40 R	73	131	4.63

From Eddleman et al. An evaluation of plot method for alpine vegetation. *Bull. Torrey Bot. Club* **91:** 446, 1964.

[a]C = circle; R = rectangle.

individual of a species can occur within the plot and all other species have no space in which to occur.

Again, note that disadvantages of the plot method to estimate density include the arbitrary definition of an individual, the different sizes of the units to be used for different life-forms, and the change in definition of an individual as an observer gains experience.

5.3.3   Strip Quadrats

A strip quadrat is a rectangular shaped quadrat that has one long dimension, length, which is exaggerated relative to its width. Some literature references make use of the term "transect" to describe the sampling unit. References also are made to uses of a "belt transect." Technically, all line transects have width and, in reality, are long, thin quadrats, or equivalently, strip quadrats. These long quadrats can be subdivided into smaller units that can then be used to study the relationship between quadrat size and density estimates. The width of the strip will usually vary between 2 and 10 m but will ultimately depend on plant diameter and spacing. Tree densities are often estimated by use of strip quadrats and are enumerated within diameter size classes. These size classes for diameters are used, especially by foresters, to estimate timber volume ready for harvest.

Any quadrat of small size can be used as a modification of the strip-quadrat method. Simply place small quadrats at predetermined intervals along a transect. The advantage of using a transect is to ensure a sample from the entire stratum or vegetation type. The size of the quadrat, which is a segment from a strip quadrat, depends on the size and spacing of trees or shrubs. Statistical analysis is then conducted as if the strip quadrat had a mean density estimated from the mean of the individual small quadrats placed along the transects. There is no valid estimate of the density variance within an individual strip quadrat unless the small units are randomly placed along the transect. This should present no problem since interest is centered on estimates obtained from a larger (strip) quadrat.

Fonda (1974) measured density, basal area, cover, and frequency in temperate, moist coniferous forests in Washington. A line of five points was placed 60 m apart, and at each point a 405-m² circular plot was used with the prism method of Lindsey et al. (1958). All trees greater than 2.5 cm dbh were counted by species and expressed as number of trees per hectare. Shrub, forb, grass, and moss growth forms may also be counted at each 1-m interval.

Woodin and Lindsey (1954) used the strip-quadrat method with a 244-m line bent at right angles at midpoint, and the strip extended 3 m on either side of the centerline. The configuration, called an "elb," was used because of the linear patterns in vegetation caused by gullies, ridges, slope bases, rock joints and crevices, or particular bedrock strata. The long length was used so that the data could be treated statistically as if they were random, even though they were not. If plots are not located randomly to begin with, then the length of extension does not compensate for the nonrandom location.

Lindsey (1955) compared the strip-quadrat method to full counts for density in several forest types. Six to 10 strip quadrats, 122 m long by 6.1 m wide, were observed at each station. Complete counts could be obtained by dividing the station into 30 m strips and counting all individuals in that strip. Trees are temporarily marked to avoid counting twice. Counts are made for all stems with dbh greater than 10 cm. Six strip quadrats are counted initially in a stand and used to estimate the number of quadrats needed to estimate density within 20% or less of the standard error and within a 67% confidence level. This criterion is applied to one or two species with moderate density, rather than the most abundant species, so that the density of less abundant species would be adequately sampled.

Ranges in sizes used for the strip quadrat include 3 × 152 m for shrubs, herbs, and large and small seedlings and 6 × 152 m for other species categories to study altitudinal variation in forest composition (Scott et al. 1964). The arm's-length rectangle method, which is really a variable-strip quadrat, can be used to obtain density estimates (Atkinson 1963). The observer paces along a predetermined compass line with outstretched arms and counts individual plants encountered. For example, Atkinson counted all canopy trees and woody plants greater than 1.8 m tall in the first 25 paces, followed by counting only understory plants between 0.3 and 1.8 m in height along the next 25 paces. The next 50 paces were uncounted, and the process was then repeated. The arm's-length rectangle method can be used to obtain a density estimate by recording numbers observed in each size class of a species or other strata such as life-form.

Note that whenever a plot method is used to obtain density estimates, the plot boundary must be accurately located. It should be obvious that the use of nonrigid quadrat sides, such as the arm's-length technique, will result in inaccurate data because of inaccurate location of the outer quadrat boundaries. Therefore, whenever any quadrat is used that has nonrigid sides, right angles of the quadrat should be determined (Fig. 5.2). The length of the arms are variable, and the guide must always be long enough to ensure that a stable position is maintained along the line.

5.3.4 Considerations

Density estimates for herbaceous vegetation have been obtained frequently by a 1 × 1-m quadrat. Dice (1948) used a 0.8-m² quadrat to determine density in a grass–herb field. Smaller units, such as the 20 × 50 cm, may be used in more dense vegetation, such as meadows, annual grasslands, alpine–arctic, tundra, and subtropical vegetation types where individuals are small and are in close proximity to one another. Hanson (1934) rec-

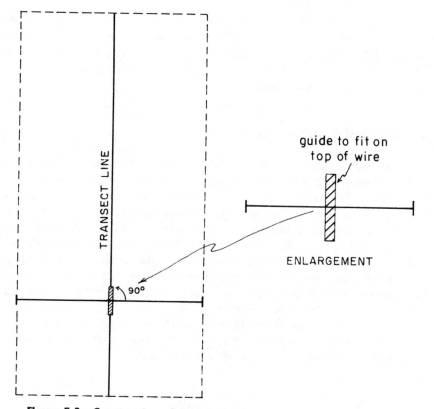

Figure 5.2 Construction of right angles for nonrigid sides of quadrats.

ommended the use of 0.1-m² plots for sampling in a mixed prairie type, while Heady (1958) used 6.45-cm² plots in the California annual type. Eddleman (1962) used a 10 × 40-cm plot to measure plant density for alpine–tundra vegetation.

Density estimates from forested areas are usually expressed as number of trees per acre or hectare. Lang et al. (1971) found 10 × 20-m quadrats for a tropical forest had the lowest required sample size for the majority of species counted. On the other hand, Bormann (1953) recommended use of 10 × 140-m plots to determine density in hardwood forests. Bourdeau (1953) used 10 × 10-m plots to determine density in a deciduous forest. These results are in contrast to those of Cain (1936), who used 50-m² plots for studies in deciduous forests. Mueller-Dombois and Ellenberg (1974) suggested the use of 10 × 10-m quadrats to determine tree density, and they further suggested a 4 × 4-m quadrat to be used for quadrat sampling of woody undergrowth up to 3 m in height.

In summary, the major problem in obtaining a reliable estimate of density is that of plot size and shape. The size, shape, and number of plots ultimately used depend on the species found in the vegetation type and their pattern of distribution. The size of plot used should be of major concern since spatial exclusion may occur for some species. In addition, the size of plot should always be sufficient to overlap small scales of patterns as indicated by patchiness in vegetation.

5.4 DISTANCE METHODS

Distance measurement techniques have been used extensively since the 1950s to estimate density. These techniques are often referred to as "variable-plot" or "plotless" methods. These are techniques that do not use rigid boundaries and are based on the concept that the number of plants per unit area can be estimated from the average distance between two plants or between a point and a plant. The distance measure usually estimates the radius of a circle that is imaginary. The distance technique may save considerable time and could even improve the accuracy of the estimate because, theoretically, no boundary errors exist (Curtis 1959, Greig-Smith 1964, Mueller-Dombois and Ellenberg 1974).

Density for plant species may be estimated by dividing the unit reference area (such as hectare $= 10,000$ m^2) by the estimated mean area of an individual plant. Mean area MA in equation form is

$$MA = \frac{1}{\text{density}}$$

That is, MA is the amount of average space (square units cm^2, 1 m^2, etc.) available to an individual plant. As seen in Figure 5.3a, the distance d extends the distance to the point or next closest plant. Thus, the distance

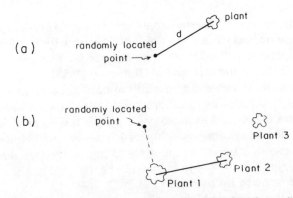

Figure 5.3 (a) Closest individual, point-to-plant method to obtain distance measure d; (b) nearest-neighbor method for measurement of plant-to-plant distance d.

measure (radius) is used to estimate area, which, in turn, is used to estimate density. Mean area is then defined as the reciprocal of density. In effect, an estimate of density is obtained from an estimate of mean area for an individual species. The remaining problem is to find a distance measure that will give the best estimate of the mean area per plant. In practice, the mean area is determined by averaging numerous distance measures for a species found in a vegetation type.

Techniques to obtain distance measures fall into one of two categories: (1) those to be used only for species that are randomly distributed (techniques include closest individual, nearest neighbor, random pairs, and the point-centered quarter) and (2) those used for species that are either randomly or nonrandomly distributed. Techniques available for use in both types of dispersion include the angle-order method, the wandering-quarter method, and the corrected-point-distance method (Batcheler 1973).

5.4.1 Random Populations

Plants and/or points must be randomly selected for measurement of distance. A purely random selection of plants involves the numbering of all plants of a species in a vegetation type and randomly selecting a subset of these numbers. Numbers in the subset represent plants from which a sample distance is measured to their nearest neighbor of the same species. If only total plant density is needed, then any species that is the nearest neighbor is measured for distance. Obviously, pure randomization is too time-consuming to be of practical use. Therefore, there are several ways of selecting a random starting point from which measurements are made. Use some process that is free of personal bias to locate points in the field. It is important to note that consequences do exist if the spatial pattern of individual plants deviates considerably from that of a random population (Persson 1971). Namely, most of the estimators for density are seriously biased. If random distributions are assumed for plants, then the following techniques can be used to obtain various distance measures.

Closest Individual Method. The closest individual method measures the distance from a randomly selected point to the nearest plant (Cottam et al. 1953, Cottam and Curtis 1956). This is the simplest method used to measure distance to estimate mean area when plants are randomly distributed over an area. However, it is also the least accurate and may give extremely biased estimates for density of a species if the species distribution, in fact, is clumped rather than random. The measure involves only the distance between a given point and an individual plant (Fig. 5.3). The equation for an estimate of density is

$$\text{Density} = \frac{\text{area}}{(2\bar{d})^2} \qquad (5.7)$$

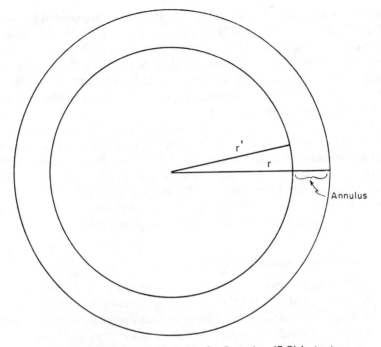

Figure 5.4 Geometric basis for Equation (5.8) in text.

where $\bar{d}$ is the distance between a point and the nearest individual plant on a unit reference area basis. For example, the area may be a hectare and the measure of $\bar{d}$, distance, uses meters to obtain the proper estimate of density per unit area.

Morisita (1954) derived the estimate for density from this technique by assuming a randomly distributed population for individual plants. The total area for vegetation is considered to be completely covered by small equal-sized quadrats and by circles of unit radius ($r = 1$) with the plot center as its center (Fig. 5.4). A Poisson series is obtained when the number of plots containing 0, 1, 2, and so on individuals is considered and if the circles are equidistant and nonoverlapping and their individual areas are small compared to the whole area. The number of individuals distributed over the whole area is assumed to be large. Then the probability of finding the closest individual in the annulus (Fig. 5.4) between distances r' and r is

$$P(r - r') = (1 - e^{-mr^2}) - (1 - e^{-mr'^2}) = (e^{-mr'^2} - e^{-mr^2}) \qquad (5.8)$$

This derives from the probability of encountering no individual plant per unit area, which is

$$P(m = 0) = e^{-m} \qquad (5.9)$$

and

$$P(m > 0) = 1 - e^{-m} \tag{5.10}$$

where m is the mean density of plants. Since area of circles depend only on values of r (the radius), Equation (5.8) is obtained. Note that r, the radius, is equal to d, the distance, sought between a point and a plant.

The mean number of individuals m present in the area A is estimated from the corresponding probability density function for r

$$f(r) = 2mre^{-mr^2} \tag{5.11}$$

and the mean is

$$\bar{r} = \frac{1}{2} \sqrt{\frac{A}{m}} \tag{5.12}$$

where $\bar{r}$ is the mean distance d between a random point and the closest individual plant. Then, from Equation (5.12), solve for m by

$$m = \frac{A}{4\bar{r}^2} = \frac{A}{(2\bar{r})^2} \tag{5.13}$$

which is identical to Equation (5.7) if $\bar{d}$ is substituted for $\bar{r}$. The probability relationship for Equation (5.7) can be used also to obtain an estimate of the density m as

$$m = \frac{1}{r^2} \log \frac{N_0 - n}{N_0}$$

where N_0 is the total number of distance measurements and n is the number of measurements shorter than r. The distribution curve for $P(r - r')$ approaches a normal curve when the square root transformation for the distance data is applied (Morisita 1954). Therefore, any data analysis on distance values d should be first transformed by taking the square root of all values for d.

Pollard (1971) presented an equation that provides for an unbiased estimate of density. Note that distances are treated as radii of circles and multiplied by pi (π) to obtain area. The closest plant to a random point method to estimate density then provides an estimate from

$$m = \frac{n - 1}{\pi \sum_{j=1}^{n} r_j^2} \tag{5.14}$$

where n is the number of randomly located points and r represents individual distances to the nearest plant from the random point.

The variance of the estimated density is

$$\text{Var}(m) = \frac{\lambda^2}{n - 2} \tag{5.15}$$

and

$$\lambda = \frac{n}{\pi \, \Sigma_{j=1}^{n} \, r_j^2} \tag{5.16}$$

To estimate density by measuring the second or third nearest individual from a random point, the following equation from Pollard (1971) will be used

$$m = \frac{K - 1}{\pi \, \Sigma_{j=1}^{n} \, r_j^2} \tag{5.17}$$

where

$$K = \sum_{j=1}^{n} k_j \tag{5.18}$$

and k_j is the order of the individual measured. That is, $k = 3$ if the third nearest plant is measured from a point. Then, if $k = 3$ at every point and 50 points are sampled, $K = 150$. The variance of m in Equation (5.17) is

$$\text{Var}(m) = \frac{\lambda^2}{K - 2} \tag{5.19}$$

where

$$\lambda = \frac{K}{(\pi \, \Sigma_{j=1}^{n} \, r_j^2)} \tag{5.20}$$

It is more difficult to measure the third nearest plant to a random point compared to selecting the nearest plant to a point. There is no doubt that more opportunity for error exists in the determinations of the third and higher-order nearest plant. Then, from the practical viewpoint, the third nearest individual plant may be the furthest one to measure from a point, if indeed it can be identified as the third one in distance. The greater the density, the more difficult it becomes to determine the order in the distance of the plant.

Nearest-Neighbor Method. The nearest-neighbor method involves a distance measure between two individual plants instead of between a point and an individual (Cottam et al. 1953, Cottam and Curtis 1956). The distance between individual plants (or culms), that are closest together near a randomly located point, is measured (Fig. 5.3b). Morisita (1954) showed that $d = \frac{1}{2} \sqrt{MA}$. Thus, $(2d)^2 = MA$, which can be interpreted as having 50% of pairs of plants occurring as each other's nearest neighbors or as duplicates. Then, 50% of all pairs occur as nearest neighbors in isolated pairs. (Also, see the discussion on the random pairs methods below.) Therefore, when individuals of a species or population occur at random, the average distance $\bar{d}$ between them is multiplied by a correction factor of 2.00, as in Equation (5.7).

However, Cottam and Curtis (1956) found that the correction factor should be 1.67 instead of 2.00 because the method does not provide a random sample of nearest-neighbor distances. In other words, 60% of the 100% sample are duplicated values instead of 50%. Thus

$$\frac{100}{60} = 1.67$$

Then, mean area per individual is less than expected from the random assumption. It should be noted, however, that the correction factor of 1.67 was determined empirically from a synthetic population, and it is unlikely that a single correction factor can be found which will always allow for the bias in the sample of distances (Pielou 1959). In any case,

$$\text{Density} = \frac{\text{area}}{(1.67\bar{d})^2} \qquad (5.21)$$

A random point is first located in the area to be measured for distances between two plants. Then the nearest plant of interest is the first plant. This plant's nearest neighbor is found, and the distance d between the two plants is measured (Fig. 5.3b).

Holgate (1964) considered the efficiency of nearest-neighbor estimators for density relative to density estimates from quadrats for random populations. As before, the nearest neighbor distances are those between an initial sample individual plant and its nearest neighbor of the same species if individual species density is sought. Let d be the distance to the rth nearest plant and $y = \pi d^2$ and $Y = \Sigma y_i$. Then, density m is estimated by

$$m = \frac{(nr - 1)}{Y} \qquad (5.22)$$

and

$$\text{Var}(m) = \frac{m^2}{(nr - 2)} \tag{5.23}$$

where n is the number of distance measurements made. If density is estimated by counting numbers of plants in randomly placed quadrats of total area A, then

$$m = \frac{Z}{A} \tag{5.24}$$

and

$$\text{Var}(m) = \frac{m}{A} \tag{5.25}$$

from knowledge of the Poisson distribution [Z = (quadrat area)(total plant count); that is, from distance measure and quadrat count; Eqs. (5.23) and (5.25), respectively]. If the two estimates are compared on a basis of sampling equal total areas, then Equation (5.25), for quadrat estimates of variance, becomes

$$\text{Var}(m) = \frac{m^2}{nr} \tag{5.26}$$

Note that this variance is smaller than the one in Equation (5.23) Therefore, the nearest-neighbor method is slightly more efficient with respect to variance than are count quadrat methods for random populations. However, this slight advantage decreases with large sample sizes (i.e., as n becomes larger) and higher-order distances, which gives larger values for r. Since more time is required as r increases, then r greater than 3 (i.e., the third nearest neighbor) may become ineffective for cost because more time is spent searching the ground for the third and fourth closest individual, and so on.

Random-Pairs Method. The random-pairs method has been widely used in forestry studies and was evaluated extensively by Cottam and associates (Cottam and Curtis 1949, Cottam et al. 1953, Cottam and Curtis 1956). This technique is similar to the nearest-neighbor method, but the random pairs method measures the distance between two individuals (Fig. 5.5). The plant [1 (i.e., plant 1 in Fig. 5.5] (a shrub, tree, or herb) nearest a sample point, which is either systematically or randomly placed, is located. The investigator then faces this plant (1) with arms spread out to both sides to form

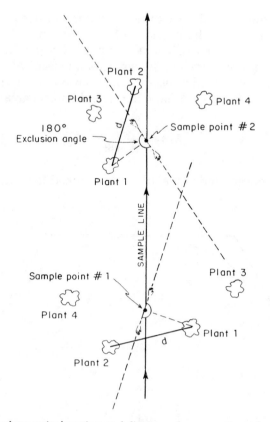

Figure 5.5 Random-pairs location and distance *d* measurement from plant to plant.

two imaginary lines. The first line is from the observer to the plant (1) nearest the sampling point, while the second line passes through the outstretched arms of the observer. The purpose of these lines is to establish a 180° exclusion angle. The second plant (2) to which the sample distance *d* is measured is the nearest plant (2) behind the outstretched arms of the observer. The mean distance $\bar{d}$ is multiplied by a correction factor of 0.87 (derived below) before squaring to obtain the mean area. Then,

$$\text{Density} = \frac{\text{area}}{(0.87\bar{d})^2} \tag{5.27}$$

Care should be taken that an individual plant is not measured twice. Otherwise, the estimate will be biased, even though the individual plants are randomly distributed.

The random-pairs technique is a reasonably accurate means of estimating frequency, density, and dominance of tree species in forests. A specific assumption of the method is that the distribution of trees or other plants deviate randomly from a hexagonal pattern in which each plant is equidistant from its neighbors. Obviously, any serious deviation from this assumption biases the estimate. The radius of an inscribed circle in a hexagon is equal to one-half the distance between one plant and any of its nearest neighbors (Fig. 5.6). First, determine the mean area *MA* from

$$MA = \frac{1}{\text{density}} \tag{5.28}$$

Since spacing between individual plants is assumed to be hexagonal (i.e.,

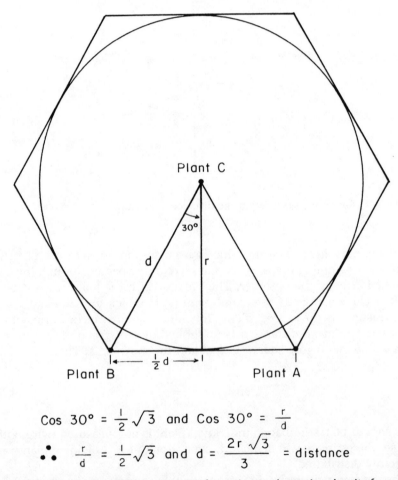

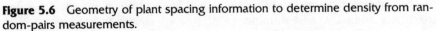

Figure 5.6 Geometry of plant spacing information to determine density from random-pairs measurements.

occurs on the corners of a hexagon and one in the center; points A, B, and C in Fig. 5.6 are examples), then one plant has a mean area of

$$1 \text{ unit of area} = \left(d \cdot \frac{3}{2\sqrt{3}} \right)^2 = [d(0.87)]^2 \tag{5.29}$$

since

$$d = \frac{2\sqrt{3}}{3}$$

from Figure 5.6 when $r = 1$. That is, the circle is a unit circle giving a radius of $r = 1.0$ unit distance.

Alternatively, from Equation (5.28), substitute

$$\text{Mean area} = \frac{1}{\text{density}} = \frac{1}{\text{numbers} \cdot \text{area}^{-1}} = \frac{\text{area}}{\text{numbers}}$$

Then

$$\text{Numbers area}^{-1} = \frac{\text{area}}{\text{mean area}}$$

Therefore, Equation (5.27), as before, is

$$\text{Density} = \frac{\text{area}}{(0.87\bar{d})^2}$$

To make an empirical test for the appropriateness of the 180° exclusion angle, randomly select the plant nearest equally spaced points on a compass line. Construct an angle with the line joining the point and the nearest plant as the bisector. Select the plant closest to the first plant, but outside this exclusion angle, as the second plant, and then measure the distance between the two plants for density calculations. Construction of this new angle is an attempt to establish an average distance between the plants rather than to obtain the distance. A 160° angle was arrived at empirically when various angles were tested against quadrat samples and 100% surveys by Cottam and Curtis (1949). Other angles might be more suitable in other vegetation types than the forested one described by Cottam (1947). Angles are estimated by closing one eye and looking over the bridge of the nose, then alternately closing the other eye and looking over the bridge, or by holding the fist of fingers at arm's length.

Cottam and Curtis (1949) made a comparison of the density estimates obtained from the random-pairs method, with a 160° exclusion angle, to

TABLE 5.2 Comparison of Random-Pairs Method and Quadrat Method

	Percent Density	
	Random Pairs	Quadrat
White oak	63.4	63.0
(*Quercus alba*)		
Black oak	25.4	28.0
(*Q. velutina*)		
Bur oak	3.0	2.8
(*Q. macrocarpa*)		
Cherry	5.2	4.5
(*Prunus serotina*)		
Hickory	0.7	0.6
(*Carya* spp.)		

From "A Method for Making Rapid Surveys of Woodlands" by G. Cottam and J. T. Curtis, *Ecology*, 1949, *49*, 101–104. Copyright © 1949 by The Ecological Society of America. Reprinted by permission.

density estimates from a 100-m² plot (10 m on a side). Table 5.2 displays selected density data in relative percent for the two estimates of density using 100 quadrats and distances d measured between 65 random pairs. As can be seen, the methods yield comparable results for the species of trees listed.

Cottam and Curtis (1956) published a correction for the various exclusion angles used in the random-pairs method. For example, they found that the 160° angle gave distance values equal to approximately 120% of the square root of the mean area *MA*. Hence, when 160° exclusion angles were used

$$MA = \left(\frac{\bar{d}}{1.20} \right)^2 \tag{5.30}$$

where $\bar{d}$ is the mean distance found between trees. Their original paper in 1949 contained the assumption of a hexagonal distribution, but later Cottam et al. (1953) studied several exclusion angles in assumed random populations of plants. The correction for a 180° angle is $0.87d$, as before. When this latter correction was applied to data from Rice and Penfound (1955), the error in the density estimate relative to a complete census was less than 2% for total number of trees (estimated 472 trees/ha vs. actual count of 464 trees/ha). Relative density estimates by individual species, however, deviated greatly from the census. There were large differences between a census and estimates by random pairs when the number of pairs was less than 30. Most species were encountered less than 30 times, the recommended minimum for a sufficient sample. On the other hand, some species might have aggregated (clumped) distributions, and then larger sample sizes are needed to obtain comparable estimates to those obtained from census data.

Point-Centered Quarter Method. The point-centered-quarter (PCQ) method involves distances that are measured from a point to the nearest plant in each of four 90° sectors around a randomly or systematically established sampling point (Fig. 5.7). The mean area occupied by a plant is determined by averaging the four distances of a number of observation points. Density is then determined by squaring the reciprocal of the average mean distance $\bar{d}$ per point. Note that no special attention is given to individual sample points with four measures of distance. One is made from each quadrant. However, as indicated later, Equation (5.31) does not provide an unbiased estimate of density, while the equation given later provides an unbiased estimate. The correction factor of one (1) is implied where

$$\text{Density} = \frac{1}{\bar{d}^2} \qquad (5.31)$$

for a unit of area. Cottam and Curtis (1956) recommended a minimum of 20 points for an adequate sample, but the sample adequacy equation given in Chapter 3 [Eq. (3.43)] should be used. The PCQ method is generally favored over the other plotless methods and has been used extensively in vegetation types throughout the world.

Mean area for plants is the square of the mean distance between points and individual shoots if individual species are disregarded. Otherwise, each species distance must be measured in each quarter at each point.

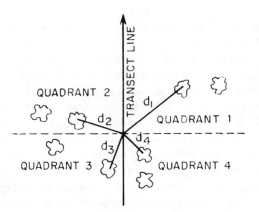

Figure 5.7 Point-centered-quarter method to measure distance *d* of plants in four quadrants from a point.

Density, then, is estimated from the reciprocal of mean area (i.e., $1/MA$) as before.

Dix (1961) applied the PCQ method successfully to grassland vegetation in western North Dakota. An individual plant was defined as an aerial shoot including its stem plus its leaves. A pin with four marks indicating the four quarters was lowered at a systematically placed point, and the distance from the point to the nearest edge of the closest shoot was measured. Risser and Zedler (1968) evaluated the PCQ method on six prairie grassland stands by comparing results with quadrat (10 × 10-cm) counts. Such a comparison should always be made because the PCQ method underestimates densities for aggregated species. This is true because there is a higher probability of a sample point occurring between, rather than within, clumps of plants. Risser and Zedler found that the PCQ method gave different results for density estimates than the quadrat method. The quadrat method tended to encounter slightly more species.

For comparison, the PCQ density estimates can be plotted against quadrat density estimates, and least-squares estimate can be made of equation coefficients for

$$y = a + bx \qquad (5.32)$$

where y is the predicted density obtained from PCQ if x (quadrat density) is used as a predictor. Coefficients a and b are the intercept and slope of the straight line, respectively. Risser and Zedler did not find differences in density estimates from the two methods when considered over all species.

Table 5.3 provides an illustration of some differences found in the PCQ and quadrat estimates of species densities. In particular, densities are underestimated by PCQ when species are aggregated as in stands 1 and 2 and are overestimated when species are approximately regularly spaced as in stand 5.

Newsome and Dix (1968) studied tree density of the Cypress Hills forests in Alberta and Saskatchewan using the PCQ method. Preliminary field tests on several stands indicated that 15 points per stand provided means for density within 5% of values obtained from 30 points. That is, only 15 points (60 distance measures) were needed to obtain an estimate within 5% of the estimate given by 30 points. Tree species were divided into three strata: trees with dbh of 94 mm or greater, saplings with dbh of 25–93 mm, and seedlings with dbh of not less than 25 mm. They compared the PCQ method with 15.2-m arm's-length quadrats described under strip-quadrat techniques. The arm's-length quadrat method was not found to be as efficient as the PCQ method, on the basis of the number of quadrats needed to obtain reliable data and difficulties encountered in dense stands. Seventy quadrats or more were needed for adequate sampling, while the 15 points were adequate for the PCQ method.

TABLE 5.3   Density Values for Some Species Within Grassland Stands as Measured by the Quadrat and PCQ Methods

	Stands					
	1		2		5	
Species	Quadrat	PCQ	Quadrat	PCQ	Quadrat	PCQ
Big bluestem (*Andropogon gerardii*)			498	298		
Little bluestem (*A. scoparius*)	188	34			1250	808
Sideoats grama (*Bouteloua curtipendula*)					116	157
Hairy grama (*B. hirsuta*)	224	62				
Bicknell sedge (*Carex bicknellii*)			270	164		
Richardson sedge (*C. richardsonii*)					182	205
Spikerush (*Eleocharis yuccifolium*)			104	39		
Ontario gayfeather (*Liatris cylindracea*)					214	205
White sweetclover (*Melilotus alba*)						
Panic grass (*Panicum depauperatum*)	30	3			64	60
Canada bluegrass (*Poa compressa*)					388	446
Kentucky bluegrass (*P. pratensis*)	378	191	354	199		
Indiangrass (*Sorghastrum nutans*)	30	34				

From "An Evaluation of the Grasslands Quarter Method" by P. G. Risser and P. H. Zedler, *Ecology*, 1968, *49*, 1006–1009. Copyright © 1968 by The Ecological Society of America. Reprinted by permission.

Heyting (1968) used the PCQ method to determine changes in vegetation density and composition in Rhodesia. Points were spaced a distance d apart on the line, and lines were $\frac{1}{2}d\sqrt{3}$ apart. Distances to the nearest living shoot are recorded in each quarter as usual. If relative density is the proportion of the total number of quarters in the sample where the species is recorded, then

$$m_R = \frac{\sum_{k=1}^{n} X}{4n} \tag{5.33}$$

The number of quarters in which the species (i) is encountered is X, and k is the number of quarters. The variance of the relative density is

$$\text{Var}(m_R) = \frac{1}{16n} \left[\sum_{k=1}^{4} (k^2 p_{ik}) - \left(\sum_{k=1}^{4} k p_{ik} \right)^2 \right] \tag{5.34}$$

since

$$\text{Var}(X) = \sum_{k=1}^{4} (k^2 p_k) - \left(\sum_{k=1}^{4} k p_i \right)^2$$

where p_k is the probability that $X = k$ for any one sample and p_i is estimated by

$$p_{ij} = \frac{N_{ij}}{N} \tag{5.35}$$

where N_{ij} is the number of times species i was encountered in j of the four possible quarters and N is the total number of measurements (number of points times 4). If distances between systematic points are greater than the diameter of the largest clumps of vegetation, and clumps are arranged somewhat randomly, then these equations may be used for systematic sampling. Variance m_R computed from systematic data might be an overestimate if, indeed, it is at all biased.

The PCQ method has been criticized because it gives biased results in clumped distributions, requires large amounts of office computations, and is sometimes inefficient in the field (Schmelz 1969). The problem with this criticism is the two to three point samples obtained for comparison. The PCQ method overestimated density in most cases by as much as 60%. Therefore, Schmelz considered the PCQ method to be insufficient for a detailed analysis of hardwood forests. As in many studies that have compared methods, no consideration was given to statistical adequacy of sampling. That is, an adequate size of sample should have been obtained to make valid comparisons.

Mark and Esler (1970) tested the accuracy of the PCQ method versus complete tallies in several forests: three *Nothofagus* forests in Fiordland and three such forests in western Otago. Complete tallies of tree stems greater than 10-cm dbh were made in 0.04-ha (100 × 4-m) plots, and PCQ data were obtained from 50 systematically placed points within each plot (25 in western Otago stands).

Mean densities for plot and PCQ method of estimating total density are given in Table 5.4. Overall density estimates obtained by PCQ appeared to be too low for some forests if plot values are considered to be the true values. Perhaps the distribution of larger canopy trees approached a ran-

TABLE 5.4 Comparison of Density Estimates Based on the PCQ Method and a Plot (0.04 ha)

	Fiordland								
	Mountain Beech Stand			Mountain–Silver Beech Stand			Silver Beech Stand		
	Plot	PCQ	% Dev	Plot	PCQ	% Dev	Plot	PCQ	% Dev
Total density (stems/ha)	622	697	+12	442	514	+16	536	504	−6
	Western Otago								
	Red Beech Stand			Mountain Beech Stand			Silver Beech Stand		
	Plot	PCQ	% Dev	Plot	PCQ	% Dev	Plot	PCQ	% Dev
Total density (stems/ha)	482	351	−27	558	573	+3	785	736	−6

From Mark and Esler. 1970. An assessment of the point-centered quarter method of plotless sampling in New Zealand forests. *N.Z. Ecol. Soc. Proc.* **17:** 106–110.

dom distribution, while smaller trees tended to cluster. This is not uncommon in plant reproduction cycles. Furthermore, large trees occupy more area. These two explanations may account for underestimates of density by the PCQ method and should be considered if the method is used.

Pollard (1971) suggested that an unbiased estimate for density could be obtained from the PCQ method from

$$m = \frac{4(4n - 1)}{(\pi \, \Sigma_{i=1}^{n} \, \Sigma_{j=1}^{4} \, r_{ij}^2)} \tag{5.36}$$

where n is the number of randomly located points and r is the individual distances to the nearest tree in a given quadrant. The variance of m is given by

$$\text{Var}(m) = \frac{\lambda^2}{4n - 2} \tag{5.37}$$

where

$$\lambda = \frac{16n}{\pi \, \Sigma_{i=1}^{n} \, \Sigma_{j=1}^{r} \, r_{ij}^2} \tag{5.38}$$

Pollard found that the PCQ method apparently has no advantage over the simple distance method (i.e., the distance to a plant from a randomly located point).

Locations of PCQ random points on steep terrain with a mosaic of communities is difficult, and density is usually underestimated because distances are overestimated. This results when the measurement tape is not perfectly straight and horizontal to make the measurement of distance. Moreover, one of a group of stems is recorded for plants. Most of these individual stems would fall within a plot and result in an underestimation by the PCQ method compared to a plot.

Angle Method. An angle method can be used to measure the closest individual in each angle sector around each sample point. One variation includes the use of a plant, rather than a point, as the center. The same equations to estimate density presented for the closest individual method above are also applicable, with the addition of an adjustment factor (Morisita 1954). If m/k individuals are found in each of k sectors of unit radius ($r = 1$) where m is the number of individuals in a given area of size π (recall that area of a circle $= \pi r^2$), then

$$\bar{r} = \sqrt{\frac{\pi}{m}} \tag{5.39}$$

and

$$\sigma^2 = \frac{4 - \pi}{m} \tag{5.40}$$

are estimates for mean density and its variance with $k = r$ since there are four sectors in the PCQ method. Generally, m/k should be substituted for m in equations for the closest individual method.

The angle method divides the area around a random sample point into k equiangle sectors (Morisita 1957), and then the estimates of density and its variance are

$$m = k \frac{kn - 1}{\sum_{i=1}^{kN} r_i^2} \tag{5.41}$$

where the summation in the denominator is over kN and

$$\sigma_m^2 = \frac{m^2}{kN - 2} \tag{5.42}$$

where N is the total number of distances measured in a sample.

Dice (1952) used spacing between individuals in a slightly different way to obtain an angle method to estimate plant density. He claimed that his method was useful to determine whether the plant distribution was aggregated. In his approach, individuals are randomly selected, and distances to the center of the nearest neighbor of the same kind of species in each of six sectors (recall assumption of the hexagon distribution for plants) surrounding that point are measured.

If no individual of the appropriate species is found in a sextant within the plot, and the vegetation is not different outside the plot, then the distance to the nearest individual outside the plot is measured. If the vegetation is different, then that sextant is omitted from further consideration. Points that are too close to the boundary to obtain data for all sextants are avoided since this will bias data toward shorter distances. Duplicate measurements arising from closeness of two points of origin and their occurrence as a nearest neighbor in one of the other sextants are also estimated.

Density is estimated from mean distance d for regular or random populations by

$$m = \frac{1.555}{d^2} \tag{5.43}$$

This equation should not be used for data that are strongly skewed to the left or right, which is a common condition in some natural populations.

5.4.2 Distance Methods for Random and Nonrandom Populations

The second group of distance measures, which may be used for randomly and nonrandomly dispersed populations, includes the order method, the angle–order method (Morisita 1957), the wandering-quarter method (Catana 1963), and the corrected-point-distance method (Batcheler 1973). The angle–order method is not to be confused with the angle method presented above.

Order Method. Morisita (1957) described an order method that requires measurement to the nth nearest individual from a random point of origin. Table 5.5 contains the moments (mean and variance μ and σ^2, respectively) from which density m can be estimated. Higher precision for an estimate is obtained by measurements to farther individuals since $(\sigma/\bar{r})^2$ decreases from the first closest individual to the second, third, and so on. A density estimate is obtained from

$$m = \frac{nN - 1}{\sum_{i=1}^{N} r_i^2} \tag{5.44}$$

TABLE 5.5 The Density Functions of Distribution, the Moments, and the Values of $\sigma/\bar{r}^2$ in the Order Method

Order'	Density Function $f(r)$	$\mu_1'(=\bar{r})$	μ_2'	σ^2	$(\sigma/\bar{r})^2$
1st closest individual	$2mre^{-mr^2}$	$\dfrac{1}{2}\sqrt{\dfrac{\pi}{m}}$	$\dfrac{1}{m}$	$.21460\,\dfrac{1}{m}$	$.8584\,\dfrac{1}{\pi}$
2nd closest individual	$2m^2r^3e^{-mr^2}$	$3\left(\dfrac{1}{2}\right)^2\sqrt{\dfrac{\pi}{m}}$	$\dfrac{2}{m}$	$.23285\,\dfrac{1}{m}$	$.4139\,\dfrac{1}{\pi}$
3rd closest individual	$\dfrac{2}{2!}m^3r^5e^{-mr^2}$	$\dfrac{3\cdot5}{2!}\left(\dfrac{1}{2}\right)^3\sqrt{\dfrac{\pi}{m}}$	$\dfrac{3}{m}$	$.23883\,\dfrac{1}{m}$	$.2717\,\dfrac{1}{\pi}$
4th closest individual	$\dfrac{2}{3!}m^4r^7e^{-mr^2}$	$\dfrac{3\cdot5\cdot7}{3!}\left(\dfrac{1}{2}\right)^4\sqrt{\dfrac{\pi}{m}}$	$\dfrac{4}{m}$	$.24170\,\dfrac{1}{m}$	$.2020\,\dfrac{1}{\pi}$

From Morisita 1954. Estimation of population density by spacing method. *Kyushu University Faculty Science Memorial Series E* **1**: 195.

similar to Equation (5.41) except the summation is over N, not kN, and k is not a product in Equation (5.44). Then

$$\sigma_m^2 = \frac{m^2}{nN-2} \tag{5.45}$$

A combination of angle and order methods (the angle–order method) is recommended for best results. Comparison of observed values of the mean and variance with theoretical ones in Table 5.5 should be made for non-random populations or when such is suspected.

Angle–Order Method. The angle–order method is based on the assumption that the area may be divided into several small fractions A_i or sectors on which the individuals will be distributed randomly or uniformly, even though individuals are distributed nonrandomly over a large area A (Morisita 1957). Density may then be estimated from the individual small fractions of the area. This approach may be applicable to a mosaic of patches of different plant densities, but probably is not as useful when clumps are each composed of relatively few individuals. Sample points are randomly or regularly placed over an area, and the area around each point is divided into equiangular sectors. The distance r to the nearest nth individual, say $n = 3$, in each of four sectors k at N points is measured (Fig. 5.8). Then, two methods of calculating the estimates for density are available. One involves the calculation of $1/r^2(r = d)$ for each sector and results in the estimates

$$m_1 = \frac{n-1}{N}\sum_{j=1}^{kN}\frac{1}{r_j^2} \tag{5.46}$$

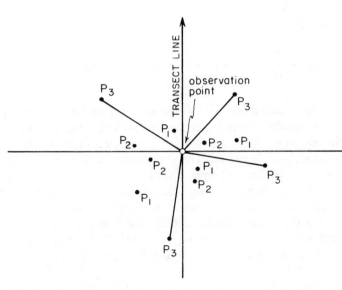

Figure 5.8 Angle–order technique where measurement of distance d is to third nearest plant in each quadrant from a point (P_i = plant order in each quadrant).

$$\sigma^2_{m_1} = \frac{1}{kN}\left[\frac{1}{A}\left(1 + \frac{1}{n-2}\right)\sum_{i=1}^{a} A_i m_i^2 - m^2\right] \qquad (5.47)$$

where m_i is the density estimate from distance measures from a small area or sector and m is the estimate of density over all points. The number of small areas (sectors) is indicated by a. The density estimate m does not depend on A_i or m_i, so these need not be known as long as individuals are distributed randomly over A_i. The density estimate m in Equation (5.46) does not involve the area A_i that is only a fraction of A (total area covered by the vegetation type). Nor is the estimate affected by m_i, the mean density of the fraction area A_i. The other method to estimate density uses

$$\sum_{j=1}^{k} r_j^2$$

for each sample point, and then the second method for estimating density is

$$m_2 = \frac{nk - 1}{N}\sum_{i=1}^{N}\frac{k}{\sum_{j=1}^{k} r_{ij}^2} \qquad (5.48)$$

and its variance is

$$\sigma_{m_2}^2 = \frac{1}{N} \left[\frac{1}{A} + \left(1 + \frac{1}{nk-2} \right) \sum_{i=1}^{a} A_i m_i^2 - m^2 \right] \qquad (5.49)$$

The first method calculates the density in each sector (small area A_i), then averages these density estimates across all sectors. The second method calculates the average density at each sample point and then averages this estimate across all sample points. The estimation method for density with greater accuracy is determined by a comparison of $\sigma_{m_1}^2$ and $\sigma_{m_2}^2$. The first method is the best if there is a significant difference in density among sectors at a sample point. The second method is more accurate if the distribution over the entire area is random and $n \geq 3$ and $k \geq 2$.

Some plant populations, however, have relatively uniform dispersal within a clump, rather than a random distribution. The sum

$$\sum_{j=1}^{k} r_j^2$$

is smaller for a uniform distribution than for a random one, and

$$\hat{m}_1 < m < \hat{m}_2$$

The best estimate of m is then obtained by

$$m = \frac{\hat{m}_1 + \hat{m}_2}{2} \qquad (5.50)$$

if a hexagonal distribution is assumed. This estimate is used for random or uniform distributions, but $\hat{m}_1$ should be used exclusively if $\hat{m}_1 > \hat{m}_2$. Note that in Equation (5.47), arithmetic considerations require that n be at least 3. It is recommended that $k = 4$ and $n = 3$ or 4. That is, use the distance to the third or fourth individual in each of four sectors.

Morisita (1957) tested the angle–order method on artificial populations and found that density estimates for regularly, randomly, and nonrandomly distributed populations were within 0.1, 1.8, and 5.0%, respectively, of actual density. In a study of Arizona desert grasslands, Strickler and Stearns (1962) found that the angle–order method estimated density more accurately than either the quadrat (quadrat size varied by species) or the PCQ methods. These results were obtained since the angle–order method is more accurate for contagiously distributed individuals. However, be aware that even quadrat estimates of density may be biased because of contagious distribution of individuals, and a specific study for quadrat size should also be made to have a valid comparison of methods.

Laycock (1965) compared the angle–order method to 0.9- and 8.9-m^2 plots to estimate density for bluebunch wheatgrass (*Agropyron spicatum*), arrowleaf balsamroot (*Balsamorhiza sagittata*), and threetip sagebrush (*Artemisia triparitita*) (Table 5.6). The distance was measured to the third closest plant in each of four quarters around the sample points.

Table 5.7 gives a comparison for the sampling precision of density estimates obtained by counts from plots and the angle–order method. A significant difference in numbers of points required for an adequate sample is seen between species in area 1 compared to those of area 2. The forb was not common on area 1 but was on area 2. So a larger variance in distances are expected that will lead to a larger sample size needed.

Lyon (1971) used the angle method to estimate density of woody plants in Idaho greater than 45 cm high. He concluded that the distance method did not obtain adequate estimates of densities on postburn communities of maple (*Acer* sp.) and snowberry (*Symphoricarpos* sp.) because of low density of shrubs. The same individual shrubs were sometimes recorded for different points and is a violation of assumptions made for use of the method. At no time should the same measures (the same plant) be repeated in any technique for measuring distance.

Wandering-Quarter Method. Catana (1963) developed the wandering-quarter method to estimate density and used distribution-free statistical methods since most plant populations are not randomly distributed. It is an adaptation of the PCQ method. Four transects are positioned where one parallel pair is perpendicular to the other parallel pair so that the lines formed a rectangle (Fig. 5.9). A directional line determined by a compass is established. The quarter is defined by the compass line and the two 45°

TABLE 5.6 Estimated Number of Plants per Hectare (in Thousands) in Two Areas as Determined by Distance in Angle–Order Method and Count in 0.9- and 8.9-m² Plots

Species	Area 1			Area 2		
	Angle–Order Method	0.9-m² Plots	8.9-m² Plots	Angle–Order Method	0.9-m² Plots	8.9-m² Plots
Bluebunch wheatgrass	52	73	43	98	83	45
Arrowleaf balsamroot	1	0.6	0.8	19	19	15
Threetip sagebrush	24	27	23	20	19	15

From Laycock. 1965. Adaptation of distance measurements for range sampling. *J. Range Manage.* **18**: 205–211.

TABLE 5.7   Comparison of Adequate Sampling Needed for Density Estimates for Angle–Order Method and 0.9- and 8.9-m² Plots for Two Areas

Species	Points or Plots Needed[a]		
	Angle–Order Method	0.9-m² Plots	8.9-m² Plots
Area 1			
Bluebunch wheatgrass	106	40	9
Arrowleaf balsamroot	210	1936	320
Threetip sagebrush	50	52	3
Points or plots sampled	30	20	10
Area 2			
Bluebunch wheatgrass	32	44	10
Arrowleaf balsamroot	38	112	71
Threetip sagebrush	45	67	20

From Laycock. 1965. Adaptation of distance measurements for range sampling. *J. Range Manage.* **18**: 205–211.

[a]Number of points or plots needed to sample within 20% of the population mean at the 95% level of confidence:

$$n = \frac{t^2 CV^2}{p^2}$$

where $t = 2.000$
          $CV =$ the coefficient of variation
            $p = 20$ (desired accuracy, %)
            $n = CV^2/100$

pie sections to either side of the line. The distance to the nearest plant within this quarter is measured. The plant then becomes the point, and the method is continued throughout the area for 25 or more distances in one compass direction. The transect line will rarely be a straight line. Therefore, the method is termed "the wandering-quarter method." The area should be tested for plant density homogeneity by systematically locating points along two perpendicular transects, measuring the distance to the closest individual within four quarters placed on each point, and noting trends in measurements. If a trend is present (e.g., distance measures increase or decrease along the transect), then heterogeneity is present.

The mean distance between individuals is determined by

$$\bar{d}_m = \frac{1}{N} \sum_{j=1}^{n} d_j \tag{5.51}$$

where d_j is the distance for jth measurement and N is total number of distances for the four transects combined. A frequency distribution of the

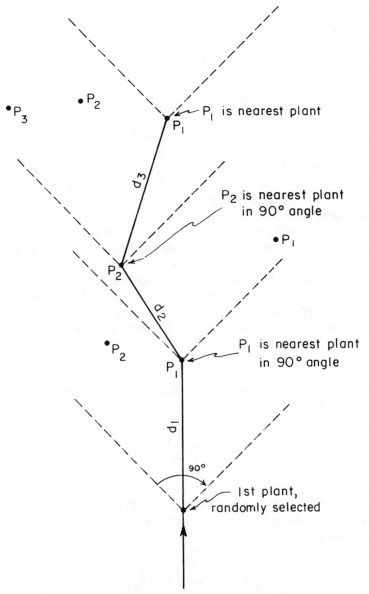

P_1 is nearest plant

P_2 is nearest plant
in 90° angle

P_1 is nearest plant
in 90° angle

90°

1st plant,
randomly selected

TRANSECT STARTING POINT

Figure 5.9 Wandering-quarter method; 90° angle always relative to the original straight line transect. Distance d measured from plant to plant.

distances can be constructed, from which the median and mode are calculated from

$$\text{Median} = L + \frac{(N/2) - s}{f} w \tag{5.52}$$

$$\text{Mode} = \text{mean} - 3(\text{mean} - \text{median}) \tag{5.53}$$

where L is lower limit of the median class, N is total frequency, s is the
partial sum up to the median class, f is the frequency in the median class,
and w is the interval width. The median will give a smaller estimate of the
variance and, thus, a better estimate of density in extremely skewed data
that result from highly aggregated populations.

To continue with the analysis, distance data are then separated into
within-clump measurements y and between-clump measurements x by
using (3 times the mode) (or 3 times the median for skewed data) as the
lower limit for the y values. No separation is obtained for randomly distributed populations because values will have a mode of 3 or less. For
random populations, and assuming that the mean distance between individuals will equal the square root of the mean area of the individuals
(Cottam et al. 1953), the mean area of the individuals is

$$MA = \bar{d}^2 \tag{5.54}$$

where MA is the mean area of individuals in terms of distance $\bar{d}$. Density
per unit area is

$$m = \frac{A}{MA} \tag{5.55}$$

where m is the density per unit of area A.

Let n_y be the number of y values and n_x the number of x values if a
separation were obtained. Then the average within-clump distance $\bar{y}$ and
between-clump distance $\bar{x}$ are

$$\bar{y} = \frac{\Sigma y}{n_y}, \quad \bar{x} = \frac{\Sigma x}{n_x}$$

The variance of the x distances is calculated in the usual manner to determine an adequate sample size since there were usually few between-clump
distances. The technique relies heavily on methods of describing characteristics of the clumps, which were fully derived by Catana (1963).

In general, the means of the within-clump distances $\bar{y}$ are determined
and are then used to obtain the within-clump area per individual W, area

of the clump C, and the mean area of the clump K. These values are then used to obtain an estimate of density.

The mean within-clump area per individual W is determined from Equation (5.54). The area of the clump C is obtained by reducing the mean within-clump distance $\bar{y}$ to unity. (All other distances are correspondingly reduced.) Then, apply correction factors to transform the wandering transects across the clumps to straight line transects, and correct these transects to the diameter of the clump. A diameter x and radius r_1 of clumps are determined from data. The number of individuals per clump N is

$$N = 3r_1^2 + 3r_1 + 1 \qquad (5.56)$$

If the mean area of the individuals within the clump W and the number of individuals per clump N are known, then the area of the clump C will be equal to the product of these two quantities

$$C = WN \qquad (5.57)$$

The actual mean diameter of the clump $\bar{z}$, assuming near circular clumps, is

$$\bar{z} = \sqrt{\frac{4C}{\pi}} \qquad (5.58)$$

Assume a random distribution for clumps, and the mean area of the clumps K is equal to the square of the sum of the mean distance between clump peripheries $\bar{x}$ and the mean clump diameter $\bar{z}$

$$K = (\bar{x} + \bar{z})^2 \qquad (5.59)$$

The number of clumps in an area is then estimated as

$$\text{Number of clumps} = \frac{A}{K} \qquad (5.60)$$

where A is the unit of area. Finally, the population density m is then estimated from the number of individuals per clump n and the number of clumps as

$$m = (\text{number of clumps}) \, N \qquad (5.61)$$

Catana (1963) noted that the number of between-clump distances is often small, and calculations should be made in order to determine whether an adequate sample size has been obtained. Again, this may be done by estimation of the variance of the x distances alone. If the standard error of the mean of these distances is less than the desired percentage of the mean $\bar{x}$, then an adequate sample has been obtained. If an inadequate sample has been taken, then the number of the distances necessary is estimated as

$$n_x = (P\ CV)^2 \tag{5.62}$$

where n_x is the number of between-clump distances required in order to have the standard error of the mean of these distances less than P, the percentage of the mean. The coefficient of variation CV is calculated from the existing sample of between-clump distances.

Several workers have reported successful results with the wandering quarter method. However, other evidence has been reported that biases may exist. For example, Catana (1963) indicated biases of low estimates for contagious populations and high estimates for regularly distributed populations.

Corrected-Point-Distance Method. Batcheler (1971) described a further modification of the PCQ method, the corrected-point-distance method, which permits sampling in nonrandom populations. Distances are measured from a sample point to the nearest member of a population, from that member to its nearest neighbor, and from that neighbor to its nearest neighbor. As in point-to-nearest neighbor measures, this measure of distance will not be biased unless the population is uniformly or contagiously distributed (Batcheler and Bell 1970, Batcheler 1973).

The procedure involves the measurement of distances obtained from systematically located points to the nearest individual r_p and from that member to its nearest neighbor r_n and from the nearest neighbor to its nearest neighbor r_m, exclusive of the member nearest to the initial point. Suppose that only distances less than R can be measured. In practice, this distance may be as little as 100 cm. Changes in values of R do not affect d for random populations except when R is small enough to exclude a given percentage (say, 50%) of the distances.

If only distances less than R are measured, the estimate of $\bar{d}$ (the mean distance) is

$$\bar{d} = \frac{n}{\pi\ (\sum_{i=1}^{n} r_i^2 + (N - n)R^2)} \tag{5.63}$$

where r_i is the distance measured between ith set of plants, n is the number of points at which a plant is found at a distance less than or equal to R, and N is the total number of observation points. Density estimates are increasingly larger since R is increased for uniform populations and is nearly twice as high when all measurements (i.e., values of R) are included. Therefore, R should be chosen to include only 50% of the distances to offset overestimates that result in uniform populations and underestimates that occur in aggregated populations. Estimates obtained in this manner are called the "50% point-distance estimate" (50% PDE).

Point distances r_p are usually larger than the joint plant-to-plant distances r_n in aggregated populations, but less than r_n in uniform cases. Hence, a function of $\Sigma r_p/\Sigma r_n$ appears to be a suitable correction factor. A linear regression of 50% PDE for population density m against $\Sigma r_p/\Sigma r_n$ was obtained by Batcheler (1971) with a correlation coefficient of -0.923, but reasonable estimates were obtained only for aggregated populations. The sign (minus) of the correlation coefficient indicates that estimates for density increases as plant-to-plant distances decrease in relation to point-to-plant distances. Then, distances are approaching a uniform distribution.

When the linear function for regression is replaced by an exponential function of r_p/r_n, it gives much better estimates of densities for all populations. Batcheler found the regression for all populations with a correlation coefficient of $-.993$ to be

$$\log m = \log d - \left(.1416 - .1613 \frac{\Sigma r_p}{\Sigma r_n} \right) \qquad (5.64)$$

Batcheler and Bell (1970) developed a model for estimating density for uniform, random, or aggregated populations.

$$m = \frac{a}{\pi \left[\Sigma_{i=1}^{a} r_i^2 + (N - a)R^2 \right]} \qquad (5.65)$$

where m is density, N is total number of sample points, R is the maximum distance over which a search is made at any point, a is the number of points at which a member is found at a distance less than or equal to R, and r_i is ith the distance measured. Although m approaches true density as R decreases, the variance will increase because fewer measurements are included. Densities in uniform populations are usually overestimated because sample points are always close to a plant. Densities in aggregated populations, on the other hand, are underestimated because distances tend to measure the distance between clusters of plants, not individual plants. These biases may be overcome by multiplication of the density estimate by the sum of the distances from the sample points to the nearest population members and dividing by the sum of the distances from those members to their nearest neighbors; in other words, $\Sigma r_p/\Sigma r_n$. This correction factor is less than unity (one) for uniform populations, and greater than unity for aggregated populations. Then the PDE has to be adjusted accordingly.

In final form, let distance be estimated as

$$\bar{d} = \frac{N}{\pi \sum_{p=1}^{N} r_p^2} \tag{5.66}$$

where symbols are as before. Then, this estimate of distance is the uncorrected estimate that is biased as a function of nonrandomness of the population. Furthermore, let

$$A_1 = \frac{1}{0.5227} \sqrt{\frac{N \sum_{p=1}^{N} r_p^2 - (\sum_{p=1}^{N} r_p)^2}{\sum_{p=1}^{N} r_p \sum_{n=1}^{N} r_n}} \tag{5.67}$$

be an index of dispersion-dependent bias of $\bar{d}$ with regard to the first order of aggregation (i.e., an index of the intensity of clumping), and

$$A_2 = \frac{1}{0.5227} \sqrt{\frac{N \sum_{p=1}^{N} r_p^2 - (\sum_{p=1}^{N} r_p)^2}{\sum_{p=1}^{N} r_p \sum_{m=1}^{N} r_n}} \tag{5.68}$$

be an index of dispersion-dependent bias of d with regard to the second order of aggregation. That is, A_2 is an index of the intensity of clumping within the clumps. Density m may then be estimated by using

$$m = \frac{m_1 + m_2}{2} \tag{5.69}$$

where

$$m_1 = \frac{\bar{d}}{3.473} (3.717)^{-A_1} \tag{5.70}$$

and

$$m_2 = \frac{\bar{d}}{3.473} (3.717)^{-A_2} \tag{5.71}$$

Systematic points may be used along random lines for increased speed, but then each line is to be used as a single observation. If density is not homogeneous along the transect, then the line should be stratified and a weighted mean computed based on density and number of sample points in each unit. Sparse populations should be sampled with a relatively small

R value and a large number of points, greater than 30, because of the distance equation [Eq. (5.66)] behavior. Then R must be increased greatly beyond the modal distance to include points in the tail that have too much influence on the density estimate. All distance measures to estimate density should be tested for acceptability for use in a vegetation type before put to general use.

5.5 VARIABLE-RADIUS METHOD

Bitterlich (1948) presented a totally different approach to estimation of density without using bounded plots. He described a variable radius or variable plot method as an efficient way to count trees in an area. Trees are counted in a circle from a central sampling point with an angle gauge (Fig. 2.10). Only trees that have a diameter that are larger than a specified angle are included in the count. The circular plot around the central sampling point has no fixed radius, but rather the radius varies with the diameter of each tree counted. That is, trees with a small diameter are included in the count only if they are close to the observer, while large-diameter trees are included at greater distances away from the observer. Because no fixed area sample is used, the method is considered to be plotless.

This method, then, because it has no fixed plot size, cannot be used directly to determine the density of a population directly (Rice and Penfound 1955). However, Rice and Penfound found that tree (or other plants) densities can be estimated from basal area data if the size distributions of the trees are known. Measurements of the diameters of each tree are made and then placed within appropriate size classes that form a frequency histogram, as described in Chapter 3. The method is very time consuming and probably should not be used unless histograms are based on several hundred tree trunk diameters. If used for herbaceous vegetation and shrubs, measurements are made on basal sizes and stem diameters, respectively.

5.6 LINE TRANSECT

The probability of a circle (a plant) of particular diameter bisected by a transect at random is proportional to an area surrounding the transect with boundary at radius r distance from the transect line and its terminal points. The expected number of contacts for all plants of the same diameter is proportional to this area and the number of individuals in the class or species.

Chords of intersection are measured along a transect, and partial chords are measured only at one end of the transect. These chord lengths C are summed. An unbiased estimate of the number of circles (i.e., density) per unit area is

$$m = \frac{\Sigma\, C\ (1/D)}{nL\ \frac{1}{4}\pi\, \Sigma\, D} \tag{5.72}$$

where n is the number of random transects and each of length L and D is the diameter of plants (circles). Since the expected value of C is $\frac{1}{4}D$,

$$m = \frac{(2/\pi)\, \Sigma\, (1/C)}{nL} \tag{5.73}$$

One modification of the above is the length of the longest chord parallel to the transect of an individual plant whose center is within a given radius of the transect. The equation for estimation of density from the longest chord parallel to the transect is

$$m = \frac{\Sigma\, (1/D)}{nL}\, \frac{1}{C} \tag{5.74}$$

or density can also be estimated from transect chords with supplementary data on longest chords by the following equation

$$m = \frac{k^2}{\bar{C}nL\, \Sigma\, D}\, \frac{AM}{HM} \tag{5.75}$$

where AM and HM are the arithmetic and harmonic means, respectively, of longest chords parallel to transects of plants intersected by the transect and k is number of plants intersected. Density estimates for ellipses instead of circles used for plant shapes are obtained by

$$m = \frac{(2/\pi)\, \Sigma\, (1/C)}{\bar{C}nL} \tag{5.76}$$

McIntyre (1953) found that density was proportional to the sum of reciprocals of chords of intersection of plants. The proportionality constant was noted to vary slightly with eccentricity of average shape as long as the individuals of plants had a relatively smooth boundary. Seedlings and single-stem herbaceous plants should be estimated separately by using quadrats because of error introduced in measuring short chords of interceptions. Strong (1966) used the line-transect method to obtain density estimates that were improved over those suggested by McIntyre. In the latter study, measurements are made perpendicular to the transect across the widest portion of each intercepted plant and labeled D for diameter.

The density estimate m then is

$$m = \left(\frac{1}{D}\right)\left(\frac{A}{T}\right)$$

(5.77)

where A is the unit area and T is the transect length. This equation includes a correction for size of plants since a plant with $D = 2$ would have twice the probability of being intercepted as a plant with $D = 1$. Then, D is the average diameter encountered along the transect. This method makes no assumptions regarding the shape of plants and does not require a subjective correction factor. The theory used to derive Equation (5.77) was based on the number of interceptions possible and their probability of being intercepted in a given area.

Perhaps a more direct way, provided from the work of Lucas and Seber (1977), to estimate density is to measure the diameter of the intercepted plant canopy of a tree, shrub, or herb. If any part of the canopy is touched by the line, the maximum diameter of that plant is measured. Then

$$m = \frac{\sum_{i=1}^{k} (1/D_i)}{L}$$

(5.78)

where D_i is the individual diameter of the ith plant, there are k plants intercepted, and L is the line length. The mean and variance are estimated from the usual equations given in Chapter 3.

5.7 INDIRECT FREQUENCY

Ashby (1935) attempted to determine the relationship between density and percentage frequency in a plant population in England. Density data were collected from two series of quadrats, 25 and 2500 cm^2. The average density was determined by counting the number of individuals, and the presence or absence of individuals in each quadrat was recorded. A value for percentage frequency corresponding to the density of every quadrat was obtained (i.e., 100 values each for frequency and density). Analysis showed that the individuals were distributed almost at random in a uniform environment, although there was a small but significant degree of aggregation of the individuals.

Recall that the relation between percentage frequency and density is given by

$$p = 1 - e^{-km}$$

(5.79)

where p is the probability (estimated by frequency f) of finding a species of density m in a quadrat of area k. Then density is estimated by

$$\log_e m = -\log_e \left(1 - \frac{f}{100}\right) \tag{5.80}$$

where m is density and f is frequency as a percentage of quadrats or other appropriate sampling units that are occupied by the species.

McGinnies (1934) applied the frequency method of Raunkiaer (1912) to semiarid conditions to derive a frequency relationship with abundance as measured by density. Most studies prior to this used an arbitrary number of quadrats to suit conditions. The number of quadrats used, controls the precision of the estimate of density since frequency is a percentage of quadrats occupied. For example, frequency can be reported only in 4% intervals with 25 quadrats, while 1% intervals are obtained with 100 quadrats.

From Equation (5.80), it is seen that frequency f is related to density by

$$f = 1 - \left(1 - \frac{1}{n}\right)^m \tag{5.81}$$

where n is the number of quadrats and m is the total number of individual plants in the large area multiplied by the fraction of the total area actually sampled. Then a weighted density estimate based on area sampled is

$$m = -\frac{\log_e (1 - f)}{\log_e [1 - (1/n)]} \tag{5.82}$$

since m has been multiplied by the fraction of the area sampled by the total number of quadrats used. Division of m by fractional area measured gives an unweighted density estimate for the total area. McGinnies conducted field trials using 25 to 200 quadrats with areas of 0.1, 0.25, 0.5, 1, 2, and 10 m². The 10-m² quadrats did not give a sufficient range of frequencies to distinguish relative densities of the more common species. The 1- and 2-m² quadrats gave the best range of frequencies for the most species, but smaller quadrats differentiated the more common species. Further studies were conducted with 0.1-m² frame placed in one corner of a 1-m² frame. Correlation coefficients for frequency versus density varied from .83 to .88.

Equations for f, using different quadrat sizes, can be solved for m by using $[1 - (1/n)]$ set to .99 since log 1.0 = 0. That is, let $1/n$ approach zero. This happens as the number of quadrats n increase and

$$m_1 = \frac{\log (1 - f)}{\log .99} \tag{5.83}$$

but based on field data

$$m_2 = 2.5m_1 \qquad \text{for 1-m}^2 \text{ series of quadrats} \qquad (5.84)$$

and

$$m_3 = 1.5m_1 \qquad \text{for 0.1-m}^2 \text{ series of quadrats} \qquad (5.85)$$

Thus, an increase in density is proportional to the increase in area of the quadrat samples. The constants, 1.5 and 2.5, were derived from theoretical curves based on a random distribution.

Comparisons of different sized quadrats are needed because species differ in their distribution in various vegetation types. Correlation for paired frequencies from the same vegetation types using different quadrat sizes are known from studies to be greater than .80. And the relation between frequency for 0.1- and 1.0-m^2 quadrats is the same as the relation of the frequency of the larger quadrat to its density divided by 10. That is

$$m = f_2 \left(\frac{a_1}{a_2}\right) \qquad \text{since} \qquad f_1 : m_1 = f_1 : f_2 \left(\frac{a_1}{a_2}\right) \qquad (5.86)$$

where f_1 and m_1 are the frequency and density of the larger quadrat, respectively, f_2 and m_2 are the same quantities for the smaller quadrat, and a_1 and a_2 are the quadrat areas.

In desert grasslands, the m/f ratio is approximately constant for the 0.1-m^2 quadrats, increases from 2.7 to 27.0 for the 1-m^2 quadrats, and is equal for the two sizes when the frequency for the larger size is near 80%. Because m/f should be as small as possible, 1-m^2 quadrats are recommended for species with f less than or equal to 80 or for a species with a density of 4 m^{-2}. If a large percentage (say, greater than 30%) of all individuals belong to species with an f greater than 80, then 0.1-m^2 quadrats might be more useful than 1-m^2 quadrats.

Myers and Chapman (1953) determined the minimal size of a sample plot and the best shape for density in a shrubland in New Zealand. Minimal area curves were constructed, and an area of 36-m^2 was considered the minimal size. Eight squares (6 × 6 m) and rectangles (24 × 1.5 m) (16:1 length:width ratio) were used to determine the density of 10 individual species. The 16:1 ratio was considered to form the best shape. Analysis of variance indicated that squares had more variation than rectangles. This finding agreed with Clapham's (1932) results that indicated rectangles with the same ratio (16:1) were less variable than squares for density estimated from a grassland. They also studied whether the logarithmic series could be applied to estimate density. This method makes use of the number of species that occur in an area as presented next.

5.8 NUMBER OF SPECIES

Some attempts have been made to estimate density from the number of species that occur in an area (Myers and Chapman 1953, Williams 1947). Use is made of the logarithmic series

$$S = \ln_e \left(1 + \frac{m}{\alpha}\right) \tag{5.87}$$

where S is total number of species, m is total number of individuals in an area, $\alpha = n_1/x$, n_1 is number of groups with one individual plant, and x is a constant less than 1.0 such that $n_1 \, x/2$ is number of groups with two plants. If a graph is drawn to relate S and $\log_e m$, a linear relationship for higher values for $\log_e m$ is seen. Observed and calculated values for m can be compared and sometimes are quite different when α is calculated from $S_{pm} - S_m = \log_e pm$. However, when α is calculated from Equation (5.87), reasonable agreement can be found between calculated and observed m values. The fact that the number of species plotted against log area curves gives a good fit to the α line, when α is derived from

$$S_{pm} - S_m = \log_e p \tag{5.88}$$

is no indication that reliable estimates for m can be obtained when one sample is p times the other. Then, the number of individuals cannot be reliably estimated from the number of species present in an area.

5.9 OTHER ESTIMATES FOR DENSITY

Cox (1976) proposed two density estimators that are unbiased for a wide range of species patterns. Most previous estimators discussed are unbiased for the spatial pattern for which they were designed but are usually biased for other patterns. The method expands the approach presented by Batcheler and Bell and discussed as a corrected point-to-distance measure. First, N random points are selected in a vegetation stand, and the distance X from this point to the center of the nearest individual plant is measured. Then, the distance Y from this plant to its nearest neighbor is measured. All measurements are then divided into two sets, A and B, where

$$A = [(X_i, Y_i) : Y_i \le 2X_i, \quad i = 1, \ldots, a)] \tag{5.89}$$

$$B = [(X_j, Y_j) : Y_j > 2X_j, \quad j = 1, \ldots, b)] \quad (a + b = N) \tag{5.90}$$

The pairs of distances are relabeled as (X_{1i}, Y_{1i}) for members of A and

(X_{2i}, Y_{2i}) for members of B. Define random variables as

$$Z_{1i}^2 = X_{1i}^2 \frac{2\pi + \sin B_{1i} - (\pi + B_{1i}) \cos B_{1i}}{\pi} \qquad (5.91)$$

for $i = 1, \ldots, a$, where $\sin(\frac{1}{2}B_{1i}) = \frac{1}{2}Y_{1i}/X_{1i}$. This equation was previously interpreted by Cox and Lewis (1976).

Let θ be the mean area (1/density), which has a maximum likelihood estimator of

$$\theta = \frac{1}{2\pi} \frac{(\Sigma_{i=1}^a Z_{1i}^2 + \Sigma_{j=1}^b Y_{2j}^2)}{N} \qquad (5.92)$$

This estimator for mean area MA is adjusted by a parameter related to the spatial pattern. The chosen parameter is $p = (a/b)$, which is the probability that $Y > 2X$. The parameter varied from 0.91 for extreme regularity, through 0.25 for a Poisson distribution, to 0 for extreme aggregation where plants formed clumps (Cox 1976).

The estimator is unbiased, theoretically. While Cox found errors of 20–33% of the population mean, and standard deviations larger than the means with a sample size of $N = 100$, the errors were not consistently over- or underestimates of known density values. Then, an estimate density is

$$m = \frac{1}{\theta} \qquad (5.93)$$

from Equation (5.79) since θ is an estimate of MA, and MA is the reciprocal of density m.

Earlier, Machin (1948) had proposed an estimator for density to be

$$m = \frac{1}{a} \log_e \frac{n}{y_0} \qquad (5.94)$$

where m is number of plants per unit area, a is the size of quadrat, n is the number of randomly located quadrats, and y_0 is number of trials (samples of size n) with no individuals of that species. A graph can be drawn to relate density and y_0 for a given quadrat size. Machin's graph was noted to be sensitive to densities from 3 to 33 m^{-2} for 0.1-m^2 size quadrats where $y_0 = 50$ samples, each of size n. Recall that samples y_0 are repeated sets of observations n.

Keuls et al. (1963) expanded Essed's (1957) density method to include more general distributions other than random. Essed's method involved distance measurement to the nth individual (same as before for angle order) where the value of density is independent of location. A consistent estimate of density by the latter method is related to the inverse of the square of the average distance to the fourth tree in a homogeneous Poisson-distributed forest. Keuls et al., on the other hand, dealt with the generalized Poisson forest, which was a locally homogeneous Poisson distribution. That is, small areas have individuals randomly distributed within, but over large areas, there are large density differences from one small area to another.

An unbiased estimator of the density m of the trees at s points within the small area is

$$m = \frac{sn - 1}{\pi} \frac{1}{\sum_{i=1}^{n} d_n^2} \tag{5.95}$$

where d_n is the distances to the nth individual at each of s points. This is identically distributed with one measurement to the snth trees. The variance for the estimate is

$$\text{Var}(m) = \frac{m^2}{n - 2} \tag{5.96}$$

Distances only to the third, fourth, or fifth tree are usually measured because of the time involved in locating more distant trees and because of the area around the point must be small to make the assumption of a local Poisson distribution for plants valid. Distances may be measured at k equally spaced intervals, and the density mean and its variance are

$$m = \frac{1}{k} \sum_{j=1}^{k} m_j \tag{5.97}$$

$$\text{Var}(m) = \frac{1}{k^2(n - 2)} \left[m^2 + \frac{1}{k} \sum_{j=1}^{k} (m_j - m)^2 \right] \tag{5.98}$$

where m_j is the jth small area and k represents the number of small areas.

DuToit Deetlefs (1953) discussed various methods of expressing density as a function of number of trees and average dbh or basal area per tree. These methods included crown size classes, tree height, and relationship between dbh and crown size using several techniques. The more important equations are given here; those of height and basal area. Density, based on height, is estimated as

$$m = \frac{10,000}{(hf)^2} \tag{5.99}$$

where m is the number of trees per hectare in a natural stand, h is the average stand height (in meters), and f is a constant that is characteristic of the species and is estimated from known densities and heights of species. The density estimate from basal area BA is

$$m = \frac{BA}{(k \text{ dbh})^2} \qquad (5.100)$$

where dbh (diameter at breast height) is average dbh (m) and BA is the basal area (m^2) per hectare; k must be calculated for known densities of a natural stand.

5.10   COMPARISONS

Cottam et al. (1953) studied density sampling characteristics of an artificial random population. Sampling was performed with square quadrats of an area equal to $\frac{1}{4}$, $\frac{1}{2}$, 1, 2, and 4 times the population mean area as determined by distance measures. Distance methods used were distance from the closest individual to a sampling point to its nearest neighbor, random pairs, and the point-centered quadrat (PCQ). The PCQ method was equivalent to a quadrat with size equal to 4 times the population mean area. The random pairs method was used with exclusion angles of 0–340°. Accuracy of relative density estimates was noted to be dependent on the number of individuals measured rather than on the method used. Angle methods can usually give better results per man hour spent in terms of sample size than can quadrat methods. This is because distances are easily measured compared to counting of individuals. Sample size, once again, is the key to obtaining reliable estimates of density regardless of method. Then, economy becomes the important criteria.

Cottam and Curtis (1956) compared four distance methods for estimating density with the quadrat method: closest individual, nearest neighbor, random pairs, and PCQ. Populations sampled were three forest communities sampled with 100 observations. A summary of their density results are reported in Table 5.8. Sample size n needed to give a standard error less than 4.6% of the mean are also given in Table 5.8. This level of precision was chosen because distances had to be squared before densities were determined. Sample size for adequate sampling of basal area was always less than that needed for adequate distance sampling.

The PCQ method required more time per point, but the reduced variability among points more than offset this time. The random-pairs method was preferred over the nearest-neighbor method because it required fewer points. The closest individual method was not recommended for distance measures to estimate density because of its variability. Cottam and Curtis thought that location of sample points by pacing probably resulted in some

TABLE 5.8   Comparison of Density Estimates from Various Methods

Method	Closest Individual	Nearest Neighbor	Random Pairs	Quarter	Quadrat	Actual
		Wingra Woods				
Trees/ha	236	234	239	251	272	257
Estimated n needed	136	99	61	38	—	—
		Peavey Falls Woods				
Trees/ha	1019	859	925	988	947	932
Estimated n needed	146	94	71	26	—	—
		Jack Pine Woods				
Trees/ha	1791	1720	1560	1834	1638	1730
Estimated n needed	121	59	61	30	—	—

From "The Use of Distance Measures in Phytosociological Sampling" by G. Cottam and J. T. Curtis, *Ecology*, 1956, *37*, 451–460. Copyright © 1956 by The Ecological Society of America. Reprinted by permission.

bias because of subconscious placement near large or unusual trees. They suggested that this would affect the closest-individual method the most because only one distance is measured in the method. That is from a point to a plant.

Strickler and Stearns (1962) compared methods for estimation of plant density. Distance methods used included closest individual, nearest neighbor, random pairs, and PCQ. The PCQ method was tested in the California annual grass type, and estimates compared to those obtained on quadrats of 6, 26, and 94 cm^2. Estimates of numbers per square m obtained by PCQ were consistently lower than quadrat counts on slope sites, while even greater differences occurred in swales. This is a typical result when shoots are counted instead of whole plants in an aggregated population, because the mean area per plant is overestimated.

Angle–order methods and PCQ were applied to desert shrubs and trees in Arizona. Either angle, order, or angle–order estimate was better than the PCQ, except for species considered to be randomly distributed. The wandering quarter method provided close estimates in all cases except in random populations with abnormally large spaces lacking individuals. Then, overestimates of density occurred for these populations.

Plot counts might be the most efficient density estimate if this were the only measurement of vegetation being obtained. However, the objectives and vegetation type should be considered before any measurement technique is decided on. For example, distance methods are the most useful when applied to one species in an autecological study. Relative densities may be useful, but absolute densities of species derived from distance measures depend on accurate estimates of total density that rely on a random distribution for species. One problem with distance methods is to find the desired individual in the dense stand where other species have the same morphological characteristics.

Beasom and Haucke (1975) compared the PCQ method, random pairs, nearest neighbor, and closest individual methods to total count methods for determination of density in coastal sand dunes vegetation of southern Texas. All methods appeared to overestimate density, but this overestimate was density independent. That is, overestimates in denser stands were no greater than in less dense stands. Densities ranged from 317 to 1275 trees/ha. Only the nearest neighbor method significantly ($p < .01$) overestimated density when values were averaged across stands. The PCQ method appeared to be the most accurate, followed by random pairs and closest individual, but these were not significantly different.

Becker and Crockett (1973) compared the modified point transect, line transect, angle–order, quarter, and quadrat methods to estimate density on tallgrass prairie in eastern Oklahoma. One objective was to determine whether methods for nonrandom populations gave better results than methods that assumed a random population when applied to a highly aggregated population. Four plots, sized 1.2×5 m, were subdivided at 20-cm intervals and systematically located. The meter line transect, PCQ, and angle–order methods were applied to all four plots, while a modified point transect and 10×10-cm open-ended quadrats were only applied to two plots. Summary estimates for density are given in Table 5.9. The PCQ underestimated aggregated species and overestimated single-stalked species. Density estimates obtained by PCQ were significantly less accurate than those obtained by other methods. If the data for the angle–order method were adjusted to remove the distances of 1 cm or less, then its

TABLE 5.9 Comparison of Total Density Values (Plants/m^2) Obtained from Some Techniques

Actual Density	Line Transect	PCQ	Quadrat	Angle–Order Method[a]		
				$\dfrac{\overline{m}_{lex}}{\pi}$	$\dfrac{\overline{m}_1}{\pi}$	$\dfrac{\overline{m}_0}{\pi}$
986.7	1099.5	422.0	1009.0	923.6	1592.4	1132.2

Becker and Crockett. 1973. Evaluation of sampling technique on tall-grass prairie. *J. Range Manage.* **26:** 61–65.

[a]$\overline{m}_{lex}/\pi$ was calculated by excluding distance measurements 1 cm or less, $\overline{m}_1/\pi$ was calculated by including all distance measurements, and $\overline{m}_0/\pi$ is the average of m_1 and m_2 divided by π.

accuracy was improved and this method was better than the line transect. The quadrat method appeared to be most accurate but was not significantly more accurate than the angle–order or the line-transect method. The methods were ranked in order of decreasing time: line transect, angle–order, PCQ, and quadrat.

Laycock and Batcheler (1975) compared closest individual, PCQ, angle–order, and corrected-point-distance methods of density determination on two tussock grassland species in New Zealand and compared these values with total count data. Two 20 × 20-m sampling areas were chosen. One area contained bare areas up to 3 m across, and the other had relatively homogeneous cover.

The areas were gridded into 121 (0.56-m²) plots where densities were obtained by counting. Interior intersections were used as sample points for the distance measures. The grass formed tussocks, while the forb formed clumps and rosettes. The closest individual and PCQ methods overestimated density by 11–13% and 27–29%, respectively, for the homogeneous grass population but underestimated the aggregated population densities. The angle–order method resulted in estimates within 8% of true density for the grass population. The corrected point distance estimate was within 10% of actual population densities in all experiments.

From a study such as that described above, it is suggested that distances to the nearest individual, its nearest neighbor, and the neighbor's nearest neighbor (corrected-point-distance method) be measured for populations within unknown distributions. Distance measurements do require more time as populations become more aggregated, and it is difficult to determine exact positions of individuals, and plot counts require less time in more aggregated populations. Distance measurements need to treat individual species separately to gain accurate information. Determination of overall density and recording species encountered to obtain relative densities results in biased estimates of density for individual species. Usually, aggregated species are encountered less often by distance measures than they should be for their actual density. Bias is also known to increase as aggregation of individuals increases.

Lyon (1968) tested 19 methods of density estimates made in a shrub community in Montana. A 152 × 215-m area was gridded into smaller squares (9.3 m²), and total counts of all individuals, taller than 15 cm and spaced 15 cm or more apart, were obtained. Plants were found to be distributed in aggregations within the small square, and randomly to regularly in the large areas.

Plants were counted in 50 randomly located quadrats. Distance methods included 60 randomly located points for PCQ, angle method, angle–order, wandering quarter, and wandering angle. Techniques were evaluated on the basis of accuracy, precision, and efficiency (Table 5.10). Precision was the estimated number of samples to estimate density within 124 plants/ha ($P = .95$) of approximately 10% of the mean.

TABLE 5.10   Comparison of Some Density Estimation Techniques

Sampling Method	Sample N	Estimated Density Plants/ha	Samples Required to Estimate Mean Density Within ±124 Plants (P = .95)
Quadrat Techniques			
Squares			
0.6 × 0.6 m	50	1660 ± 1227	17,926
3.1 × 3.1 m	50	1593 ± 239	681
6.1 × 6.1 m	50	1134 ± 134	213
Rectangles			
0.3 × 1.2 m	49	564 ± 565	3,721
1.5 × 6.1 m	49	1264 ± 185	401
3.1 × 12.2 m	49	1129 ± 129	195
Distance Techniques			
PCQ	60	596 ± 79	89
Angle method	60	600 ± 79	89
Angle–order	60	635 ± 55	44
Morisita's angle–order	60	981	
Catana's wandering quarter	240	1029	
Wandering angle	240	1150 ± 107	653

From Lyon. 1968. An evaluation of density sampling methods in a shrub community. *J. Range Manage.* **21:** 16–20.

Squares, rectangles, and wandering-quarter methods resulted in sample means within two standard errors of the true density value. Distance measurements from random points were more precise than plant-to-plant or quadrat estimates, but distance methods underestimated the mean by 33%. Lyon concluded that complete enumeration is the best method because the most time-efficient methods produced biased results. He did not consider the data distribution for the estimates of count or distances. Yet, the sample adequacy equation is based on the assumption of normality of data. If results of any comparison for measurements are to be acceptable, information must be on a statistically reliable basis.

Hutchings and Morris (1959) compared nearest-neighbor, closest-individual, and PCQ methods with total plant counts on desert shrubs in western Utah. Close agreement among methods occurred only on random plants. Individual species were usually aggregated in mixed stands, but plants without regard to species were randomly dispersed. It is not unusual to note this result. That is, individual species occur in aggregated patterns while the overall population of plants occurs as random.

Shanks (1954) compared distance measures to estimate density by the random pairs method (160° exclusion angle) with conventional plot sampling in the Great Smoky Mountains. Density estimates for the 400-m^2, 100-m^2, and random-pairs method were found to be 665, 922, and 541 trees/ha, respectively, in a spruce–fir forest. All estimates were found to be within acceptable ranges statistically and were, therefore, not significantly different.

Singh and Chalam (1936) used variance methods to determine the relative efficiency of quadrats and strips in obtaining counts of individuals by species. Quadrats were 5 cm on a side and divided by 1-mm wires into 625 small squares, each with a side of 0.2 m. Accuracy of quadrat estimates was 4.5 times that of strip estimates, as determined by variance methods. This efficiency was maintained only up to a particular level of density. If the density for a species was less than 20% of the total vegetation, then the strip method was less variable by a 4 : 1 factor. Strips were 25 × 1 small squares, while quadrat data consisted of a grid of 5 × 5 small squares. Strips were more accurate in the less dense, random vegetation than in the more dense, nonrandom vegetation. Since minor species were seldom of importance, the quadrat methods were considered best.

5.11   CONSIDERATIONS

Brown (1954) stated that in order to describe the ecological characteristics of a species or a community, the attributes of frequency, number, area covered, and weight must be assessed. Density, then, is a frequently assessed parameter that is usually readily obtainable and easily understood. Problems of determining an accurate density estimate exist, however. Such problems include difficulties in determining individual plants, problems in size and shape of quadrat and boundary errors, and unbiased estimates from plotless methods. Density cannot be singularly used as a measure of dominance, but rather must be used in conjunction with other vegetation characteristics such as frequency, cover, and biomass.

The methods that are most often used and are most accurate in determining the density estimate are the quadrat and plotless methods. Strickler and Stearns (1962) noted that when the only objective in the study is to estimate density, quadrat counts may be more efficient than some distance measures. The size of quadrat used to estimate density in herbaceous and shrubby communities is dependent on the purpose of study and the community structure. Shrubs and small trees are often recorded in 16-, 10-, or 4-m^2 quadrats, and very few studies have been conducted in woody communities to determine the accuracy and efficiency of various quadrats. It appears, however, that visual estimates of density may be easier and more accurate than many sampling methods (Lyon 1968).

Herbaceous communities are most frequently sampled by 1- or 0.1-m^2 quadrats. Earlier studies in forested communities utilized 10 × 10-m quadrats. More recently, however, plotless methods have become more widely utilized in forested communities. Generally, then, the size of the quadrat ultimately used is dependent on the community structure, plant morphology, and the purpose of the study.

Distance measurement techniques are among the simplest to use, provided the population consists of discrete individual plants. Quadrat boundaries are unnecessary, and quadrat size automatically is related to the population density under consideration (Laycock and Batcheler 1975). Generally, distance measures are most useful when applied to one species in autecological studies, but these measures can be used to study an entire community. The number of samples needed, however, to adequately sample the least abundant members of a community may preclude this type of sampling. Sometimes such community studies do not require an absolute determination of number of individuals by species. Instead, a determination of relative density may be used to compare species composition between plant communities. Relative values, however, are dependent on the estimated density of all species within a community and may, therefore, be biased and misleading. When measurement of density is carried out within very dense or very sparse communities, the use of distance measures may be inefficient because considerable time may be spent on searching for the required individual plant. This problem may be more serious when measurements are made to the third or more distant plant, such as in Morisita's (1957) angle–order method.

If plotless methods are employed, distribution patterns must be assessed before the sampling method is chosen. For randomly dispersed individuals within a population, the PCQ method is the most reliable. Most populations are not randomly dispersed, but rather are contagiously dispersed. For nonrandomly dispersed individuals, the angle–order method appears to be the most reliable. The corrected-point-distance method (Batcheler 1973) appears to show some promise for providing accurate density estimates. It should be noted, however, that more extensive studies need to be conducted to determine the efficiency and accuracy of plotless sampling methods under field conditions. As Persson (1971) noted, if the spatial pattern of the individuals deviates considerably from that of a random population, most of the estimators may give seriously biased results.

Plotless methods require less time and may be more efficient for random populations. Randomness of plants may not always be the case, however, and appropriate techniques should be used in nonrandom distribution patterns. In fact quadrat methods may integrate variations in pattern and departures from randomness better than plotless methods and may, therefore, be more accurate to estimate density. In some vegetation types, however, quadrat methods may be inefficient and give inaccurate estimates of density. A detailed study is needed to determine the efficiency and accuracy of all sampling methods in a particular vegetation type before more specific recommendations can be made regarding which set of methods are appropriate.

The sample size needed is defined as the number of observations and is often determined arbitrarily. An adequate number of samples must be taken in order to obtain a reliable estimate of the true mean. An equation to determine an adequate size was given in Chapter 3 [Eq. (3.43).]. Generally, the number of quadrats counted for density is a function of the variation among individual samples. In other words, the greater the variation, the greater the number of quadrats needed. There appears to be two major ways that sample size can be determined for density: the first is a function of the area to be sampled, and the second relates to the number of individuals counted per quadrat or point.

One may set a standard of 5 or 10% sampling intensity for an area. For example, an area of a particular vegetation type covers 10 ha and employs 5000 quadrats of 1 m². Alternatively, this area could be sampled by fifty 10 × 10-m plots.

Optimum sample size to minimize variance is not a function of the area sampled, but rather is a function of the number of samples containing individuals of a species. This is related to the spacing of individuals. Species that are in close proximity may be sampled with fewer quadrats or distance measures than species which are widely spaced.

For a population with a random distribution, the sample size may be statistically determined by use of the ratio of the standard error of the mean to the mean. However, Greig-Smith (1964) also stated that this relationship is of little practical value because few populations are randomly distributed. He suggested that for nonrandom populations, the percentage standard error of the mean can be calculated and then tabulated values can be used for a given mean range to obtain sample size.

Kershaw (1964) suggested a practical guide for estimating the adequacy of the sample size. An approximation of the mean variance may be accomplished by calculating successive means. Sampling should be terminated at the point at which additional quadrats do not significantly affect the mean of the more important species. After an initial large variation, the curve (number of quadrats vs mean variance) becomes asymptotic, indicating less variation between quadrats.

5.12 BIBLIOGRAPHY

Ashby, E. 1935. The quantitative analysis of vegetation. *Ann. Bot.* **49:** 779–802.

Atkinson, I. A. E. 1963. Some methods for studying the effects of goats in forests. *N. Z. J. Bot.* **1:** 405–409.

Bartlett, M. S. 1948. Determination of plant densities. *Nature* **162:** 621.

Batcheler, C. L. 1971. Estimation of density from a sample of joint point and nearest-neighbor distances. *Ecology* **52:** 703–709.

———. 1973. Estimating density and dispersion from truncated or restricted point-point distance nearest neighbor distances. *Proc. N. Z. Ecol. Soc.* **20:** 131–147.

Batcheler, C. L., and D. J. Bell. 1970. Experiments in estimating density from joint point- and nearest-neighbor distance samples. *Proc. N. Z. Ecol. Soc.* **17:** 111–117.

Beasom, S. L., and H. Haucke. 1975. A comparison of four distance sampling techniques in South Texas live oak mottes. *J. Range Manage.* **28:** 142–144.

Becker, D. A., and J. J. Crockett. 1973. Evaluation of sampling techniques on tallgrass prairie. *J. Range Manage.* **26:** 61–65.

Bitterlich, W. 1948. Die Wenkelzahlprobe. *Allg. Forst-U. Hulzw. Ztg.* **59:** 4–5.

Blackman, G. E. 1935. A study by statistical methods of the distribution of species in grassland associations. *Ann. Bot.* **49:** 749–777.

Bormann, F. H. 1953. The statistical efficiency of sample plot size and shape in forest ecology. *Ecology* **34:** 474–487.

Bourdeau, P. F. 1953. A test of random versus systematic ecological sampling. *Ecology* **34:** 499–512.

Braun-Blanquet, J. 1932. *Plant Sociology.* (Translated by G. D. Fuller and H. S. Conard.) McGraw-Hill, New York, 439 pp.

Brown, D. 1954. *Methods of Surveying and Measuring Vegetation.* Bulletin no. 42, Commonwealth Bureau of Pastures and Field Crops, Hurley, Berkshire, 223 pp.

Cain, S. A. 1936. The composition and structure of an oak woods, Cold Springs Harbor, Long Island, with special attention to sampling methods. *Am. Midl. Nat.* **17:** 725–740.

Cain, S. A., and G. M. Castro. 1959. *Manual of Vegetation Analysis.* Harper & Bros., New York, 325 pp.

Carpenter, J. R. 1938. *An Ecological Glossary.* Oklahoma University Press, Norman, OK, cited by C. W. Cook (chairman), 1962. *Range Research: Basic Problems and Techniques.* National Academy of Sciences–National Research Council, Publication No. 890. Washington, DC, 341 pp.

Catana, A. J. 1963. The wandering quarter method of estimating population density. *Ecology* **44:** 349–360.

Clapham, A. R. 1932. The form of the observational unit in quantitative ecology. *J. Ecol.* **20:** 192–197.

Committee on Nomenclature of the Ecological Society of America. 1952. Cited by C. W. Cook (chairman), 1962. *Range Research: Basic Problems and Techniques.* National Academy of Sciences–National Research Council, Publication No. 890. Washington, DC, 341 p.

Cook, C. W., chairman. 1962. *Range Research: Basic Problems and Techniques.* National Academy of Sciences–National Research Council. Publication No. 890. Washington, DC, 341 pp.

Cottam, G. 1947. A point method for making rapid surveys of woodlands. *Bull. Ecol. Soc. Am.* **28:** 60.

Cottam, G., and J. T. Curtis. 1949. A method for making rapid surveys of woodlands by means of randomly selected trees. *Ecology* **30:** 101–104.

Cottam, G., and J. T. Curtis. 1956. The use of distance measures in phytosociological sampling. *Ecology* **37:** 451–460.

Cottam, G., J. T. Curtis, and B. W. Hale. 1953. Some sampling characteristics of a population of randomly dispersed individuals. *Ecology* **34:** 741–757.

Cox, T. F. 1976. The robust estimation of the density of a forest stand using a new conditioned distance method. *Biometrika* **63**: 493–499.

Cox, T. F., and T. Lewis. 1976. A conditioned distance ratio method for analyzing spatial patterns. *Biometrika* **63**: 483–491.

Curtis, J. T., 1959. *The Vegetation of Wisconsin. An Ordination of Plant Communities.* University of Wisconsin Press, Madison, 657 pp.

Daubenmire, R. 1959. A canopy-coverage method of vegetation analysis. *Northwest Sci.* **33**: 43–64.

———. 1968. *Plant Communities: A Textbook of Plant Synecology.* Harper & Row, New York, 300 pp.

Dayton, W. A. 1931. *Glossary of Botanical Terms Commonly Used in Range Research.* USDA Miscellaneous Publication No. 110, 40 pp.

Dice, L. R. 1948. Relationship between index and population density. *Ecology* **29**: 389–391.

———. 1952. Measure of the spacing between individuals within a population. *Contrib. Lab. Vertebr. Biol., Univ. Mich.* **55**: 1–23.

Dix, R. 1961. An application of the point-centered quarter method to the sampling of grassland vegetation. *J. Range Manage.* **14**: 63–69.

DuToit Deetlefs, P. P. 1953. Means of expressing and regulating density in forest stands. *S. Afr. Forestry Assoc. J.* **23**: 1–12.

Eddleman, L. E. 1962. Evaluation of sampling methods for alpine tundra. M.S. Thesis, Colorado State University, Ft. Collins, CO, 105 pp.

Eddleman, L. E., E. E. Remmenga, and R. T. Ward. 1964. An evaluation of plot methods for alpine vegetation. *Bull. Torrey Bot. Club* **91**: 439–450.

Essed, F. E. 1957. Estimation of standing timber. *Medel. Van de Land. te Wagen., Nederland* **57**(5): 1–60.

Evans, F. C. 1952. The influences of size of quadrat on the distributional patterns of plant populations. *Contrib. Lab. Vertebr. Biol.* **54**: 1–15.

Fonda, R. W. 1974. Forest succession in relation to river terrace development in Olympic National Park, Washington. *Ecology* **55**: 927–942.

Fracker, S. B., and H. A. Brischle. 1944. Measuring the local distribution of ribes. *Ecology* **25**: 283–303.

Goodall, D. W. 1952. Quantitative aspects in plant distribution. *Biol. Rev.* **27**: 194–242.

Greig-Smith, P. 1964. *Quantitative Plant Ecology*, 2nd ed. Butterworths, London, 256 pp.

Hanson, H. C. 1934. A comparison of methods of botanical analysis of the native prairie in western North Dakota. USDA *J. Agric. Res.* **49**: 815–842.

Harris, T. H. 1941. The sampling of ribes populations in blister rust control work. *J. Forestry* **39**: 316–323.

Heady, H. F. 1958. Vegetational changes in the California annual type. *Ecology* **39**: 402–416.

Heyting, A. 1968. Discussion and development of the point-centered quarter method of sampling grassland vegetation. *J. Range Manage.* **21**: 370–380.

Holgate, P. 1964. The efficiency of nearest neighbor estimators. *Biometrics* **20**: 647–649.

Hope-Simpson, J. F. 1940. On the errors in the ordinary use of subjective frequency estimations in grassland. *J. Ecol.* **28:** 193–209.

Hutchings, S. S., and M. J. Morris. 1959. Use of distance measurements for determining plant density in semidesert vegetation. (Abstract.) *Ninth International Botanical Congress, Montreal, Proceedings,* vol. 2, p. 174.

Kershaw, K. A. 1964. *Quantitative and Dynamic Ecology.* Edward Arnold, London, 183 pp.

Keuls, M., H. J. Over, and C. T. DeWit. 1963. The distance method for estimating densities. *Stat. Neerl.* **17:** 71–91.

Lang, G. E., D. H. Knight, and D. A. Anderson. 1971. Sampling the density of tree species with quadrats in a species rich tropical forest. *Forestry Sci.* **17:** 395–400.

Laycock, W. A. 1965. Adaptation of distance measurements for range sampling. *J. Range Manage.* **18:** 205–211.

Laycock, W. A., and C. L. Batcheler. 1975. Comparison of distance-measurement techniques for sampling tussock grassland species in New Zealand. *J. Range Manage.* **28:** 235–239.

Lewis, S. M. 1975. Robust estimation of density for a two-dimensional point process. *Biometrika* **62:** 519–521.

Lindsey, A. A. 1955. Testing the line-strip method against tallies in diverse forest types. *Ecology* **36:** 485–495.

———. 1956. Sampling methods and community attributes in forest ecology. *Forestry Sci.* **2:** 287–296.

Lindsey, A. A., J. D. Barton, Jr., and S. R. Miles. 1958. Field efficiencies of forest sampling methods. *Ecology* **39:** 428–444.

Lucas, H. A. and G. A. F. Seber. 1977. Estimating coverage and particle density using the line intercept method. *Biometrika* **64:** 618–622.

Lyon, L. J. 1968. An evaluation of density sampling methods in a shrub community. *J. Range Manage.* **21:** 16–20.

———. 1971. *Vegetal Development Following Prescribed Burning of Douglas-Fir in South-Central Idaho.* USDA Forest Service Research Paper INT-105. 30 pp.

Machin, E. J. 1948. Determination of plant densities. *Nature* **162:** 257.

Mark, A. F., and A. E. Esler. 1970. An assessment of the point-centered quarter method of plotless sampling in some New Zealand forests. *N. Z. Ecol. Soc. Proc.* **17:** 106–110.

McGinnies, W. G. 1934. The relation between frequency index and abundance as applied to plant populations in a semiarid region. *Ecology* **15:** 263–282.

McIntyre, G. A. 1953. Estimation of plant density using line transects. *J. Ecol.* **41:** 319–330.

Morisita, M. 1954. Estimation of population density by spacing method. *Kyushu Univ. Fac. Sci. Mem. Ser. E* **1:** 187–197.

———. 1957. A new method for the estimation of density by the spacing method applicable to non-randomly distributed populations (translation by USDA, Forest Service. 1960.). *Physiol. Ecol.* **7:** 134–144.

Mueller-Dombois, D., and H. Ellenberg. 1974. *Aims and Methods of Vegetation Ecology.* Wiley, New York, 547 pp.

Myers, E., and V. J. Chapman. 1953. Statistical analysis applied to a vegetation type in New Zealand. *Ecology* **34:** 175–185.

Newsome, R. D., and R. L. Dix. 1968. The forests of the Cypress Hills, Alberta and Saskatchewan, Canada. *Am. Midl. Nat.* **80:** 118–185.

Persson, O. 1971. The robustness of estimating density by distance measurements, pp. 175–187. In *Statistical Ecology,* vol. 2, *Sampling and Modeling Biological Populations and Population Dynamics,* G. P. Patil, E. C. Pielou, and W. E. Waters (eds.). Pennsylvania State University Press, University Park, PA.

Pielou, E. C. 1959. The use of point-to-plant distances in the study of the pattern of plant distributions. *J. Ecol.* **47:** 607–613.

Pieper, R. D. 1973. *Measurement Techniques for Herbaceous and Shrubby Vegetation.* New Mexico State University Bookstore.

Pollard, J. H. 1971. On distance estimators of density in randomly distributed forests. *Biometrics* **27:** 991–1002.

Pound, P., and F. E. Clements. 1898. A method of determining the abundance of secondary species. *Minn. Bot. Stud.* **2:** 19–24.

Raunkiaer, C. 1912. Measuring apparatus for statistical investigations of plant formations. *Svensk. Bot. Tidsskr.* **33:** 45–48.

Rice, E. L., and W. T. Penfound. 1955. An evaluation of the variable-radius and paired-tree methods in the blackjack-post oak forest. *Ecology* **36:** 315–320.

Risser, P. G., and P. H. Zedler. 1968. An evaluation of the grasslands quarter method. *Ecology* **49:** 1006–1009.

Robinson, D. H. 1933. III. The percentage frequency method. *Agric. Progr.* **10:** 233–234.

Schmelz, D. 1969. Testing the quarter method against full tallies in old-growth forests. *Indiana Acad. Sci. Proc.* **79:** 138.

Scott, G. A. M., A. F. Mark, and F. R. Sanderson. 1964. Altitudinal variation in forest composition near Lake Hankinson, Fiordland. *N. Z. J. Bot.* **2:** 310–323.

Shanks, R. E. 1954. Plotless sampling trials in Appalachian forest types. *Ecology* **35:** 237–244.

Singh, B. N., and G. V. Chalam. 1936. Unit of quantitative study of weed flora on arable lands. *Am. Soc. Agron. J.* **28:** 556–561.

Smith, A. D. 1944. A study of the reliability of range vegetation estimates. *Ecology* **25:** 441–448.

Stearns, F. W. 1959. Floristic composition as measured by plant number, frequency of occurrence, and plant cover, pp. 84–95. In *Techniques and Methods of Measuring Understory Vegetation.* (Proceedings of a symposium.) Southern Forest Experiment Station and Southeast Forest Experiment Station.

Strickler, G. S., and F. W. Stearns. 1962. The determination of plant density, pp. 30–40. In *Range Research Methods. A Symposium.* (Denver, CO.) USDA Forest Service Miscellaneous Publication No. 940, 172 pp.

Strong, C. W. 1966. An improved method of obtaining density from line-transect data. *Ecology* **47:** 311–313.

Van Dyne, G. M., W. G. Vogel, and H. G. Fisser. 1963. Influence of small plot size and shape on range herbage production estimates. *Ecology* **44:** 746–759.

Walker, B. H. 1970. An evaluation of eight methods of botanical analysis on grasslands in Rhodesia. *J. Appl. Ecol.* **7:** 403–416.

Whitman, W. C., and E. I. Siggeirsson. 1954. Comparison of line interception and point contact methods in the analysis of mixed grass range vegetation. *Ecology* **35:** 431–436.

Williams, C. B. 1947. The logarithmic series and the comparison of island flora. *Proceedings of Linnaen Society London* **158:** 104–108.

Woodin, H. E., and A. A. Lindsey. 1954. Juniper–pinyon east of the Continental Divide, as analyzed by the linestrip method. *Ecology* **35:** 473–489.

6 BIOMASS

All biological activities of plants, animals, and human beings are dependent on the energy of gross primary productivity. "Primary production" is defined as the energy fixed by plants, and it is perhaps the most fundamental characteristic of an ecosystem. Measurements of primary production are necessary for proper understanding of ecosystem dynamics. Vegetation composition, based on dry weight, is one of the best indicators of species importance within a plant community (Daubenmire 1968). This chapter deals with measurements of primary production in terrestrial ecosystems. This vegetation measure is referred to as "biomass" in the discussion that follows.

6.1 HERBACEOUS BIOMASS

The best and probably the most commonly used method for measurement of herbaceous production is to clip or harvest total standing biomass. Except for small experimental plots, total harvest of vegetation in a large area is neither feasible nor possible. Therefore, various sampling techniques and methodologies have been developed to obtain estimates of herbaceous biomass. These methods can be grouped into three categories: direct methods, indirect methods, and a combination of direct and indirect methods. Selection of any one method depends on its merits and the objectives of vegetation measurements.

6.1.1 Harvesting

There are a variety of implements and methods for harvesting herbaceous production; for example, sheep shears, grass and hedge shears, sickles, hand lawnmowers equipped with grass catchers, power mowers, mowing machines, and power clippers. The choice depends largely on plot size, topography, and growth characteristics of the plants. However, in native vegetation the herbaceous vegetation is commonly clipped with grass shears and browse production by pruning shears. Shears are sometimes equipped with welded, upturned edges to catch clippings from low-growing plants that cannot be held with hands. Grass shears are useful for clipping very small areas such as in cages, quadrats, and nursery rows. Lawnmowers are useful only for herbage under 15 cm tall. An ordinary mowing machine is not suitable for cutting taller vegetation (>25 cm). Clipping height above ground level depends on the objectives of the study.

Weighing and Drying Harvested Material. The weight of live plant materials includes the inter- and intracellular water and any external moisture from vapor condensation, precipitation, and so forth. Therefore, the weight of freshly harvested plant materials is highly variable and depends on the moisture status of plant and atmospheric conditions. Then, for meaningful interpretation of results, biomass is expressed in terms of oven- or air-dry weight. In large-scale surveys, it is not feasible to keep all the harvested materials to determine the air- or oven-dry weights. Harvested materials are weighed fresh, or "green," in the field and recorded. A few samples of each species or species groups are dried. The ratio of dry to fresh weight is used to convert biomass data of respective species to dry weight. All herbage from sample quadrats may either be weighed together or separated into major species or species groups. Some grassland investigators separate two or three dominant grasses, lump the remaining grasses into a group, referred to as "other grasses." Forbs, shrubs, and trees can be treated similarly. Samples are to be dried within 24 hr to keep fermentation and respiration losses to a minimum.

If samples of biomass are large and cannot be oven-dried quickly, then samples may be dried in a heated room. Herbage samples are usually oven-dried at 60°C, but temperatures between 92 and 105°C are also used. If a chemical analysis is to be conducted on the vegetative material, then the lower temperature will prevent a reduction or loss of chemical constituents. A slight loss of ammonia may occur when young plant material is placed in an oven at 95–100°C. Losses also occur in carbohydrates. A constant ratio between green and air-dry weights will not occur for composite samples made up of several species because species composition differs among sampling units and each species has a different moisture percentage. So a sample of each species should be dried separately, then weights combined to obtain total biomass estimates from the vegetation type.

Laboratory Estimation of Species Composition. An investigator may be interested in finding the contribution that different species or life-forms make to biomass production. But it may not be feasible to separate species in the field. Several methods are used to estimate species composition in the laboratory from harvested material. The most accurate of all biomass sampling methods is separation of harvested material, by hand, into individual species and each species is weighed. Laboratory separation is possible only if plants are harvested in one piece. This requirement is necessary because it is difficult to distinguish between detached parts of plants of different species, especially leaves of some native grasses. After drying, plants become brittle, and hand separation becomes difficult. The problem is overcome to an extent by a visual estimation of percentage composition of all dried samples. Then a few samples are hand-separated and species are weighed separately to check the estimates. Estimates are then adjusted by ratio or regression techniques.

The point sampling method for cover described in Chapter 4 can be used in the laboratory to analyze clipped samples from the field for species composition. Heady and Torrell (1959) used the point-sampling procedure to develop a technique for laboratory separation of clipped materials. The cross-hair in one eyepiece of a binocular scope is used as a sample point to identify plant material. Heady and Torrell found that a sample of 400 hits was necessary to obtain sample means for the major species within 10% of the true mean with a 95% confidence level. (See section 3.4.2)

Heady and Van Dyne (1965) suggested a method for predicting weight composition of species from point samples on clipped herbage. Ten plots, 30 cm on a side, are clipped and samples are composited by date. Ten subsamples are analyzed by 50 points taken under a binocular microscope. A regression equation is used to predict percentage composition by weight from composition determined by the laboratory point method.

6.1.2 Measurement of Biomass

Direct Method. Harvesting or clipping is probably the most common direct method to estimate herbaceous biomass production. A plot of known dimensions is placed on a land area. Vegetation biomass is then harvested from the three dimensional volume of the quadrat (height × width × length). Plant biomass that is not rooted in the quadrat but occupies space in the volume is harvested while portions of plants rooted in the quadrat that do not occupy the quadrat volume are not harvested (Fig. 2.16). Vegetation samples are clipped and weighed by species, aggregates of species or life-form categories and are standardized to oven- or air-dry weight. Materials such as current year's growth, all vegetation, or only palatable species for grazing animals may be harvested at various levels of intensity (i.e., the height to which vegetation is clipped). While height of clipping above ground level depends on the objectives of data collection, it is best to clip all vegetation to ground level. Then, amount of material available for forage by herbivore class can be partitioned by plant species.

The two main objectives of harvesting herbaceous biomass are to determine the amount of forage available for grazing by herbivores of all sizes, or to evaluate habitat conditions for vegetation. The advantage of harvesting is that data can be utilized for a variety of purposes. In particular, harvesting provides:

1. Availability of long-term records of vegetation production.
2. Variability in the contribution of individual species to total biomass production over time.

3. The relation of total plant yield of biomass to other factors such as soil moisture, plant cover, frequency, and density.

Sampling procedures often contain subjective decisions about inclusion of litter, certain vegetation parts such as floral structures, and standing dead material. The observer must be consistent about decisions made and must maintain a constant clipped height across all plots. Improper species identification will always be one of the most common errors made in determination of biomass production on an individual species basis. Means and variances for estimates of biomass obtained by harvesting are calculated by usual methods presented in Chapter 3.

Weight Estimates and Double Sampling. In this method a number of randomized observations are taken by visual estimation of biomass weight, and, in addition, biomass is clipped at a relatively small number of observation points taken at random from a large sample of visually estimated points. Two sets of data then become available: (1) a large sample that contains only observations by the visual method and (2) within the large sample, a small sample that contains clipped weights of biomass in addition to the estimated weights. The small sample data set is used to calculate a regression to show the relation of clipped plant weight y to the estimated weight observed by visual estimation x. Estimates of x for the large sample are then used to obtain predicted biomass weights y. The unit clipped may be plots, branches, whole plants, or parts.

That is, an estimate of the clipped mean $\overline{Y}$ over all units, had all plots or units been clipped instead of visually estimated, is

$$\overline{Y} = \overline{y} + b(\overline{x}' - \overline{x}) \tag{6.1}$$

where $\overline{y}$ is the sample clipped mean, $\overline{x}'$ and $\overline{x}$ are the means of the weights x_i in the indirect (guesses) and direct (clipped) samples, respectively, and b is the coefficient for least squares regression of $\overline{Y}$ on the difference $(\overline{x}' - \overline{x})$. The variance of $\overline{Y}$ (the corrected mean estimate of biomass from double sampling) is

$$v(\overline{Y}) = \frac{s_{y\cdot x}^2}{n} + \frac{s_y^2 - s_{y\cdot x}^2}{n'} \tag{6.2}$$

where $s_{y\cdot x}^2$ is the residual sum of squares divided by the number of double sample degrees of freedom $(n - 1)$. That is

$$s_{y\cdot x}^2 = \frac{\Sigma^n \, [y_i - (a + bx_i)]^2}{n - 1} \quad \text{and} \quad s_y^2 = \frac{\Sigma_{i=1}^n \, (y_i - \overline{y})^2}{n - 1}$$

where y_i is an individual unit of weight clipped and $\bar{y}$ is the mean for clipped weight. Say, for example, the units are plots. The x_i is the individual visual estimate of the individual clipped weight y_i. Then, the variance $v(\bar{Y})$ in Equation (6.2) is minimized by selection of n (the number of plots to be double sampled) and n' (the number of plots to be visually estimated only). This selection should always include costs to obtain both n and n'. Then, total cost C for double sampling is minimized by solving for a minimum C value when

$$C = nc_n + n'c_{n'} \qquad (6.3)$$

where c_n and $c_{n'}$ are costs associated with a plot that is both visually estimated and clipped (n) and a plot that is visually estimated only (n'). Cochran (1977) shows that Equation (6.3) and

$$\frac{n}{\sqrt{V_n C_{n'}}} = \frac{n'}{\sqrt{V_{n'} C_n}} \qquad (6.4)$$

can be used to determine n, and n' as an optimum combination. As an example, let the variance for clipped plots V_n be 36, and cost C_n equal unity and the variance for a regression double-sample estimate $V_{n'}$ be 169 with a cost $C_{n'}$ of 20% of clipped plots. That is, it costs 5 times more to clip than to visually estimate a plot. Equation (6.4) is solved by

$$\frac{n}{\sqrt{36(0.2)}} = \frac{n'}{\sqrt{169(1)}} \quad \text{or} \quad \frac{n}{n'} = \frac{\sqrt{7.2}}{\sqrt{169}} = \frac{2.7}{13} = 0.21$$

Then, Equation (6.3) is used to find n and n'. That is, if 100% is available for total costs C and $C_{n'}$ is one-fifth the cost of C_n, substitution into Equation (6.3) gives

$$100 = n + 0.2n'$$

Dividing through by n' gives

$$\frac{100}{n'} = \frac{n}{n'} + \frac{0.2n'}{n'} = 0.21 + 0.2$$

or

$$n' = \frac{100}{0.41} = 244 \quad \text{and} \quad n = (244)(0.21) = 51$$

A ratio of 4.8 (244/51) or 5:1 is then needed to estimate total biomass within the specific costs and variances expected. The ratio is also determined from $n/n' = 0.21$ and $1/0.21 = 4.8$ as before.

If Equation (6.1) is a close fit (i.e., if visual estimates are close to clipped values), then this ratio will decrease. For instance, if the variance from the regression estimate $V_{n'}$ is 72 instead of 169 and the cost ratio remains the same, then

$$\frac{n}{n'} = \frac{\sqrt{7.2}}{\sqrt{72}} = \frac{2.7}{8.5} = 0.32$$

and the ratio is 3:1. So an increase in ability to guess close to clipped weights of biomass reduces the need to clip since both provide essentially the same information. Another interpretation is that as one misses the clipped values by visual estimates, the variance from regression $V_{n'}$ increases and the ratio of n' to n increases because more visual estimates are needed to adequately estimate the variance of biomass by visual estimates. However, a combination of variance estimates from clipped and visual estimates will usually result in a more efficient sampling estimate. This is especially true when $1/n$ is small (i.e., the reciprocal of the number of clipped plots is small), which implies that a large number of plots would have to be clipped to permit adequate estimation of the mean and its variance for biomass unless double sampling is used. Since an adequate number n for double samples is needed to estimate the regression coefficients, a minimum of 25–30 such observations should be made for large areas (several hundred hectares) of homogeneous vegetation composition, that is, a vegetation type. Then, the ratio is used to determine the number n' of plots to be visually estimated. For the first example, a ratio of 5:1 was needed, and if n must be at least 25, then $n' = 125$. That is, 25 are both clipped and visually estimated for biomass.

Double sampling then enables one to make more precise estimates of the average biomass weight. Since both large (visually estimated) and small (clipped and visually estimated) samples are taken by randomization, an error variance can be calculated for the regression estimator and also for estimated plant biomass.

Ahmed and Bonham (1982) described a procedure for optimum allocation in double sampling for several species at once. Multivariate (more than one variable considered at a time) double sampling is beneficial when biomass is considered for several species separately or for classes of species (forbs, grasses; perennials, annuals; etc.). The estimates for individual species mean and variance are obtained from Equations (6.1) and (6.2), respectively. However, determination of the optimum ratio of visually estimated to visually estimated and clipped simultaneously for several species depends on individual species variances and associated variances from regression equations.

Equation (6.2) is used to find a constraint needed for each species. A small preliminary double sample will provide these estimates for each species (or class) of interest. That is, the solution for the optimum ratio is bounded by the variance of the clipped values because clipping is assumed to be more accurate than guessing. Then, Equation (6.2) is written as an inequality

$$\frac{s_{y \cdot x}^2}{n} + \frac{s_y^2 - s_{y \cdot x}^2}{n'} \leq V(\overline{Y}) \tag{6.5}$$

or $V(\overline{Y})$ can be replaced by another representation of precision estimate such as from previous experience, data from the literature, or a bound on the error of the estimate (say, 2 times s_e, the standard error of the estimated means). A series of these inequalities exists for species; one inequality for each species or class of biomass. Optimization techniques are used for an optimum n and n' based on sample estimates of $s_{y \cdot x}^2$, s_y^2, and $V(\overline{Y})$.

6.1.3 Nondestructive Methods

Measurements of vegetation biomass at a given point in time by harvesting techniques have some limitations. For example, some ecological studies warrant repeated measurements, within the same growing season, of permanently located sampling units, and this is not possible with a direct harvest method. Plant production of biomass may be taken by harvesting, but then other data such as seed-set date or seed production cannot be recorded. In situations like these, the only alternative is to select a suitable nondestructive method for estimating biomass productions. It was previously noted that harvesting is time consuming and expensive. Therefore, indirect methods are useful since they eliminate or reduce clipping and can be used to obtain a measure of biomass. These methods use ratio or regression estimators wherein certain easily measured vegetation characteristics are mathematically correlated to biomass production. Vegetation characteristics, such as leaf length, stem length, number of stems, crown diameter, basal diameter, and cover, all have a degree of correlation with biomass.

Regression equations are generally of the form

$$y = a + b[f(x)] \tag{6.6}$$

where y represents the biomass estimate for each species, a is a constant, b is the slope of the line, and $f(x)$ is a function of the measured plant characteristic. The $f(x)$ may be log x or other relationships that need to be determined for any characteristic such as cover for each species. Equations are developed for individual species. However, data can be combined from different vegetation areas in the same locale to develop the equation for each species. Data collection procedures must be kept as consistent as possible with respect to both biomass production and its correlated characteristic. Otherwise, variability associated with data collection is increased and the equation is less accurate for the estimate of biomass.

Capacitance Meter. Capacitance meters of the type described in Chapter 2 have been used extensively to determine the amount of vegetation biomass standing per unit area. This is essentially a double sampling technique that employs an electronic capacitance meter in place of visual estimation. To obtain an unbiased estimate, a number of randomized observations are taken by recording the capacitance meter value and, in addition, biomass is clipped at a few of these randomly selected sampling points. The two sets of data are used to solve a regression equation that gives the relationship of clipped biomass y to the capacitance meter value x. The values of x for the large sample are then adjusted to predict mean biomass weight $\bar{y}$.

Currie et al. (1973) tested the multiprobe meter and found that exclusion of very large amounts of standing dead organic matter or very succulent plants improves estimates of biomass production. However, the single-probe capacitance meter (Vickery et al. 1980) is insensitive to plant moisture level and presence of dead or senescent plant materials. Vickery et al. were able to account for most of the variance in pasture biomass. Toledo et al. (1980) calibrated their capacitance meter on pure stands of grass and clover, as well as a mixture. The capacitance meter reading, visual estimates of height, stand density, and species percentage data were used to develop prediction equations. In general, the meter reading correlated better with dry weight than with fresh weight.

Michell and Large (1983) calculated a regression for a single-probe capacitance meter to estimate biomass and found a change in the regression coefficient during the spring and another change to a constant level over the summer. This indicates that the capacitance meter does not measure dry weight of biomass directly, but rather, some factor that is related to herbage mass only within a given time of the growing season.

Rising-Plate–Disk Meters. The rising-plate and disk pasture meters described in Chapter 2 are widely used to measure herbaceous biomass. These meters have the advantage of being simple and accurate compared to other nondestructive methods such as capacitance meters and visual estimates. Earle and McGowan (1979) developed an Ellinbank pasture meter (EPM), an automated version of the Massey grass meter (Holmes 1974). They evaluated the EPM and found it to be more accurate than visual estimation. As use of this instrument requires little training, more measurements can be made in the same amount of time as that required for visual estimates.

Two major sources of error in biomass estimation are sampling error caused by within plot variability, and the problem of achieving a constant height of cutting. Michell (1982) found that the rising-plate meter (RPM) overcomes both of these problems. He used the RPM in a double sampling technique format and found a strong relationship existed between meter reading and herbage mass. He also found that variable readings occurred over the summer compared to those obtained during winter months.

Both the RPM and single probe capacitance meters give data that are more closely related to dry weight of the biomass than to green weight. Neither meter gives accurate results in vegetation containing large amounts of dead vegetation material. Stockdale and Kelly (1984) found the single-probe meter to be better than the RPM for biomass estimation when trampling was not a factor.

Spectral Reflectance. Spectral responses of objects indicate the patterns of reflected or emitted energy that is derived from absorbed solar energy. Because plant species have different spectral responses according to biomass concentrations, it is possible to obtain estimates of biomass from these responses. Specific bands must be matched to various levels of standing vegetation biomass. Once this is accomplished, spectral analysis can be used to measure and monitor the plant biomass of an area. Radiometers are used in field measurements of biomass when non-destructive estimation of biomass is needed.

Pearson and Miller (1972a) developed a method using a radiometer to estimate biomass of a shortgrass prairie site. They used a 0.25 m² plot, obtained spectroreflectance curves, clipped small portions of the plot, and again obtained spectral readings. The process was continued until the plot was clipped to ground level. Naturally, curves varied according to amounts of green vegetation present in the plot, and curves were predictable. In particular, it was noted that an inverse linear relationship existed between total grass biomass and spectral reflectance at a band of 0.68 μm ($R^2 = 0.70$). However, when all plant biomass was included, a direct linear relationship was found at a band of 0.78 μm ($R^2 = 0.84$). All measurements were obtained with a portable radiometer placed approximately 1.0 m above the plot.

Ratios of spectral bands are also used to predict green standing biomass of vegetation. Ratios from .78/.68 μm to .800/.675 μm have been used successfully ($R^2 > .90$) to estimate green biomass (Pearson and Miller 1972a, Waller et al. 1981, Boutton and Tieszen 1983). These workers have also noted that time of day affects the precision of biomass estimates because of changes in amount of sunlight falling on the plots. Tucker (1980), in particular, noted that readings for spectral reflectance must be made in direct sunlight and within 1 to 3 hours of noontime. Otherwise, variability caused by the solar zenith angle limits precision of biomass estimates obtained from radiometers. Significant errors also accrue when green vegetation is less than 30% of the total vegetation biomass in the area. It follows that the method is more reliable during the active growth of vegetation. On the other hand, the use of radiometers over large areas are advantageous because measurements can be obtained in a matter of seconds from either high altitude aircraft or spacecraft.

Growth Form and Production. Some vegetation ecologists have hypothesized that within a given species, volume–height or height–weight distribution is more or less constant. In other words, a species growth form is relatively constant. However, height–weight distribution of biomass of a species may vary because of major differences in amount of leaf material, for example, in a drought year. If no variations in growth form occurred, then it would be easy to determine volume or weight of plants from their height. A number of attempts have been made to correlate height of plants with their weight or volume. However, growth forms of species vary so much according to weather conditions, stand density, defoliation, available mineral nutrients, and other site factors that it is difficult to use height to predict weight or volume of plants. Plants grow taller in more favorable years or in areas with high densities and have a more cylindrical or a conical shape than under other conditions.

Several workers have studied the height–volume and biomass distribution in range grasses. The relationship between height and herbage volume (or air-dry weight) shows that in most species more volume is concentrated near ground level. As height of plants increase, biomass volume in grasses usually decreases (Crafts 1938, Clark 1945, Heady 1950). Growth form for the same species varies markedly between plants from different years, from different elevational zones, and from different sites (Table 6.1).

Even when heights are the same, there are still differences in weight distributions of plants. Because of this, models must be developed on a height–class basis. That is, variability still occurs between years and between sites within years, and differences between sites will usually be greater than those between years.

Kelly (1958) measured leaf and shoot height to predict production. The effect of time within a growing season affects the relationship between height and biomass of species. Then, a separate term in the equation is needed to account for seasonal differences. For example, the equation for predicting weight of biomass from plant height is

$$y = a + bx \tag{6.7}$$

where y represents biomass amount, a is a constant, and b is the rate of change in biomass amount per unit of change in height x. If seasonal effects exist, then Equation (6.7) includes another term for this effect z and its associated rate of change coefficient (c). That is

$$y = a + bx + cz \tag{6.8}$$

TABLE 6.1 Percentage of Biomass Removed at Different Percentage Heights Above Ground Level for Grasses Collected in the Same Year from the Oakbrush (*Quercus*), Aspen (*Populus*)–Fir (*Abies*), and Spruce (*Picea*)–Fir (*Abies*) Zones

Height Percent	(*Bromus marginatus*) Mountain Brome			(*Agropyron trachycaulum*) Slender Wheatgrass			(*Poa pratensis*) Kentucky Bluegrass		
	Oak	Aspen	Spruce	Oak	Aspen	Spruce	Oak	Aspen	Spruce
100	0	0	0	0	0	0	0	0	0
95	—	—	2	1	1	1	1	1	1
90	1	1	6	3	3	3	3	2	4
85	2	—	9	5	5	5	4	3	9
80	3	2	13	7	7	7	7	4	14
75	5	3	16	10	9	9	9	6	18
70	7	4	20	12	12	11	11	8	22
65	10	5	24	15	15	14	13	10	26
60	13	7	28	18	17	18	15	12	29
55	17	9	32	21	21	21	18	15	33
50	21	12	37	25	25	25	21	18	36
45	26	15	41	29	29	30	24	21	39
40	32	19	45	34	35	33	28	25	42
35	39	24	50	39	39	39	33	28	46
30	46	30	56	45	45	45	38	33	51
25	55	37	64	51	52	52	44	38	56
20	63	46	69	59	61	60	52	46	61
15	72	63	76	69	68	69	62	56	70
10	82	71	83	78	82	79	74	69	82
5	92	86	93	89	90	90	87	84	91
0	100	100	100	00	100	100	100	100	100

From Clark. 1945. Variability in growth characteristics of forage plants on summer range in central Utah. *J. Forestry* **43**: 273–283.

These equations apply to any situation where variables, other than height or season, are used to predict biomass amount of a species or group of species. Equation constants and coefficients are estimated from least squares procedures used in regression analyses.

Cover Method. Some studies indicate that cover is a suitable measure for prediction of biomass production of some species (Lundblad 1937, Payne 1974, Anderson and Kothmann 1982). Therefore, use of cover values as an estimate of herbage production can save a great deal of field and laboratory time. Cover–weight relationships are developed by measuring cover and fresh or oven-dry weights of species in a number of randomly located quadrats. The relationship between cover and oven-dry weight of a species is much more easily obtained than that of cover and fresh weight. A simple linear relationship often exists for the former relationship. The data are then used to compute cover–weight relationships for each species through regression analysis.

A high, positive correlation between ground cover and vegetation production often exists. Lundblad (1937) estimated weight percentages of plant composition in Scandinavian countries and considered cover as a more satisfactory variable to predict weight than the weight-guess method. Before cover is used, studies should be made in the vegetation types of interest to test the cover–weight relationships by developing appropriate regression equations.

Photographs of vegetation may be used for measuring vegetation cover. Cover is then used to predict biomass. Wimbush et al. (1967) used vertical color stereophotography to record cover changes in alpine flora. Wells (1971) modified the technique of Wimbush et al. and used two 35-mm cameras fired simultaneously to take stereophotographs at approximately chest height. The transparencies are viewed directly by transmitted light under a zoom stereoscope, and plant cover is measured by point counts made on photographs with the aid of a counter. The photographic approach to measure cover can expedite the procedure to develop cover–weight equations to estimate biomass. A photograph of a quadrat is taken before clipping and weighing the vegetation. Photographic cover and oven-dry weights are then used to solve regression equations. Thus, cover is a useful method for prediction of oven-dry weights of species, but not fresh field weight.

Density Method. Several studies have been made to predict herbaceous production from plant density, but results have not been satisfactory. All plants growing within a plot are counted to obtain an estimate of plant density. Biomass is then calculated by multiplication of the average plant weight by the average density. The density–weight method may give estimates of herbaceous production that are several magnitudes of those estimated by clipped plots. This is due, in part, to the bias introduced by the nonrandom distribution of plants and considering only "rooted" plants within plots.

Point Method. The point-frame method is extensively used for measurement of vegetation cover. If at each sampling point, species biomass is also determined simultaneously, then a regression equation can be solved to estimate biomass by point sampling. After the pins in a point frame have been lowered and all hits recorded, a quadrat is lowered over the point frame and all vegetation is clipped, dried, and weighed. Hughes (1962) estimated herbage production by use of an inclined point frame that had ten pins and a 0.9-m^2 quadrat. Both plant cover and frequency data recorded by the point method gave reliable estimates of biomass production.

Biomass Prediction from Multiple Factors. A single factor such as plant height or cover may not be reliable as a predictor of herbage yield. In fact, as previously mentioned [Eq. (6.8)], a combination of variables is usually a better predictor of biomass production than a single variable. That is, more than one variable is used in combination to obtain estimates of biomass. Variables can be added, divided, or multiplied in order to find an acceptable equation. Hickey (1961) found that basal diameter, compressed crown diameter, and compressed leaf length, all combined, gave best estimates of production. A prediction model, using only compressed crown diameter and compressed leaf length, also had a high multiple correlation with biomass. Pasto et al. (1957) used cover and height to estimate pasture herbage production and found high multiple correlation coefficients for cover and height.

Indirect estimates of standing crop for alpine species were studied by Blankenship and Smith (1966) in Wyoming. Stepwise regression analysis showed that four variables highly correlated with standing crop were foliage cover, number of stems, stem length, and leaf length. However, cost analysis showed that measurement of foliage and basal cover by the point-sampling method was the most costly of the independent variables used. Yet, omission of foliage and basal cover, from analysis of data, lowered correlation coefficient values.

Height and basal area of aerial shoots was used by Mall and Tugnawat (1973) to develop regression equations to estimate biomass production of plants in a grassland ecosystem. Mall and Billore (1974) used simple, partial, and multiple correlation coefficients to show that dry-matter production was linearly related to density and basal area, which indicated that these two variables could be used as a nondestructive index to predict biomass production.

Moszynska (1970) described a method to estimate green biomass production of the herb layer in a forest. This method is based on density D and the average current growth increment by individual species G_i. The equation

$$P = G_i D \tag{6.9}$$

is used to estimate production, where P is an estimate of production of an individual species. The current increment of annual and most perennial plant species is determined by clipping, oven-drying, and weighing. This equation assumes no error in the estimation of production.

Biomass yield can be estimated (Teare and Mott 1965) by regression equations based on leaf-area index (LAI) and longest point of the plant (LP). The mathematical product of LAI and LP is often the only important term in the equation. However, another component consists of leaf and stem material, so both the product LAI × LP and LP alone may be used in the regression equation.

Andariese and Covington (1986) estimated biomass from logarithmic regression equations to relate above-ground biomass to basal area, height, number of seed heads, and some site factors. Logarithmic equations are discussed further in Section 6.3 and following where forest biomass is discussed. Plant basal area proved to be the best predictor of plant biomass for each species. The ability to predict plant biomass from basal area has advantage over some other indirect methods, such as the weight-estimate method, because personal bias is reduced.

Meteorological Data Method. Meteorological variables such as precipitation and temperature have a direct effect on plant growth. Therefore, various attempts have been made to develop models to predict herbage biomass production from precipitation and other metereological data. Wight et al. (1984) used the ERHYM (Ekalaka rangeland hydrology and yield model) (Wight and Neff 1983) biomass production model to relate soil water and climatic parameters to plant growth. The model utilizes soil moisture, precipitation, mean air temperature, and solar radiation as input data to calculate the ratio of actual transpiration T to potential transpiration T_p as a yield index. Two-thirds of the field-measured biomass yields were within one standard deviation of the forecast yields for April, May, and June. Total biomass yield is often closely related with total precipitation in a given month. Estimates may be improved by adding temperature data.

Wisiol (1984) chose six equations (Table 6.2) to estimate grasslands yield for specific years for a variety of sites. The climatic and biomass production data used to test the models came from 28 locations in North America, Asia, Africa, Australia, New Zealand, and Europe. The smallest error was obtained with the Sahel–Sudan equation, but each equation estimated yields within 5% or less at times. These findings suggest that models based on climatic data should only be used in the area for which they are originally developed.

Dry-Weight–Rank Method. The dry-weight–rank (DWR) method for botanical analysis of pasture was first described by 't Mannetje and Haydock (1963) and later refined by Jones and Hargreaves (1979) and Sandland et al. (1982). In this technique, a quadrat is placed at random and all plant species present are recorded. Quadrat size used is not critical as long as it is sufficiently large to include at least three species in most located quadrats. In subtropical pastures, quadrats of 4, 9, 25, and 40 dm² have been used successfully. The three most abundant species in the quadrat are given a ranking of 1, 2, and 3 in terms of dry weight. A rank of 1 indicates most abundant. The procedure is repeated for 50–100 quadrats per pasture or vegetation type. The data are tabulated to give proportion of quadrats in which each species occurred in first, second, and third place. These proportions are multiplied by empirically developed factors. 't Mannetje and Haydock used 70.19, 21.08, and 8.73 as multipliers for first, second, and third rank, respectively, for improved pasture vegetation. The product of

TABLE 6.2 Mathematical Models for Predicting Grazing Land Yield from Climatic Variables

Source	Equation[a]	Climate Variables[b]
Czarnowski (1973)	$y = 0.176(V)(L)(1 = e^{-P/PE})$	P/PE = precipitation ÷ potential evapotranspiration V = saturated vapor pressure, mmHg, for mean temperature T of growing season (monthly $T > 3°C$) L = length growing season (hr daylight)
Leith (1973)	y = minimum of (Y_1, Y_2) $Y_1 = 15000(1 - e^{-0.000664P})$ $Y_2 = 15000/$ $(1 + e^{1.315-0.119T})$	P = precipitation (mm) T = mean temperature (°C)
Leith and Box (1972)	$y = 15000(1 - e^{-0.0009695(AE-20)})$	AE = actual evapotranspiration (mm)
Le Houerou and Hoste (1977) (Mediterranean model)	$y = 3.89(P^{1.09})$	P = precipitation (mm)
Le Houerou and Hoste (1977) (Sahel–Sudan model)	$y = 2.643(P^{1.001})$	P = precipitation (mm)
Rosenzweig (1968)	$y = 10(10^{1.66(\log_{10}AE-1)})$	AE = actual evapotranspiration (mm)

From Wisiol. 1984. Estimating grazing land yield from commonly available data. *J. Range Manage.* **37(5)**: 473.

[a]y = annual aboveground dry matter yield (kg/ha); e = base of natural logarithms, 2.718. Form, units, and symbols are those allowing easy comparison. Czarnowski, Leith, and Leith–Box estimates were intended for total versus above-ground yield and are shown halved here. Rosenzweig model was published in logarithmic form: $\log y = 1 + 1.66(\log AE = 1)$, where y was in grams per square meter.

[b]Annual totals except for temperature (mean).

rank and multiplier are then added to give the dry weight percent of the species.

The major limitation of the method is that most abundant or first-rank species can comprise no more than 70.2% of species makeup on a dry-weight basis. Jones and Hargreaves modified the DWR method to correct limitations of the method. One limitation of the method is that biased results occur in pastures where there is a consistent relation between quadrat dry-matter production and the order in which species are ranked for production. To solve this problem, Jones and Hargreaves proposed a weight adjustment that is based on pasture production in quadrats estimated for DWR. The weighting factor used is dry-matter production in each quadrat times the rank multiplier. This adjustment technique was found to eliminate bias.

The second limitation of the DWR method occurs in pastures heavily dominated by one species. For example, in a quadrat where species A contributes 90% of the total production, species B contributes 8%, and species C contributes 2%, normal use of the DWR method would rank species A at 70%, species B at 21%, and species C at 9%. With a cumulative ranking modification, species A could be ranked in both first and second place and B in third place. This would allocate 91% to A (70% + 21%) and 9% to species B. This procedure provides a reliable estimate of composition without a change in the standard multipliers. Therefore, this modification extends the usefulness of the DWR method.

Similarly, multipliers can be empirically developed for other vegetation types, but the method would lose its simplicity. Jones and Hargreaves (1979) and Gillen and Smith (1986) found that the empirical multipliers could be used satisfactorily in ecosystems other than the one in which they were originally computed.

The coefficients can also be predicted from the general occurrence of geometric series relationship between abundance and rank of species using either an estimate of the percentage contribution of the first ranked species or from the ratio of the contribution of two species (Scott 1986). A comparison of the geometric series and DWR estimates for different ranked species for a range of vegetation diversities is given in Table 6.3. The geometric ranking procedure allows development of ranks for whatever number of species seems appropriate and eliminate the need to assign more than one rank to a species.

Species composition may be calculated by either one of the following equations (Gillen and Smith 1986).

$$C_i = \frac{\sum_{j=1}^{3} (r_{ij})(M_j)}{N} \tag{6.10}$$

where C_i = percent composition of species i
r = number of plots in which species i received rank j
M_j = multiplier for rank j
N = total number of plots

Then

$$WC_i = \frac{\sum_{j=1}^{3} (W_{ij})(M_j)}{W_T} \tag{6.11}$$

TABLE 6.3 Comparison of Geometric Series and DWR Estimates for Different Ranked Species for a Range of Vegetation Diversities

	Geometric Series						DWR		
	1	2	3	4	5	Rest	1	2	3
Empirical									
't Mannetje and Haydock (1963)							.702	.211	.087
Jones and Hargreaves (1979)							.705	.238	.057
Estimated from Geometric Series for Different k *Values*									
.32 (P5/P1 = .01)	.680	.218	.070	.022	.007	.003	.703	.225	.072
.1	.900	.090	.009	.001	.000	.000	.901	.090	.009
.2	.800	.160	.032	.006	.001	.001	.806	.161	.032
.4	.600	.240	.096	.038	.015	.011	.641	.256	.094
.5	.500	.250	.125	.063	.031	.031	.571	.286	.143
							.730	.199	.071

From Scott, D. 1986. Dry-weight-rank method of botanical analysis. *Grass and Forage Science* p. 320. Reprinted by permission of Blackwell Scientific Publications Ltd.

where WC_i = weighted percent composition of species i
W_{ij} = summed weight of all plots in which species i received rank j
W_T = summed weight of all plots

Equation (6.10) assumes equal biomass across all plots. However, vegetation biomass seldom is uniformly distributed. Equation (6.11) is used to estimate weighted species composition. In this case, total biomass in each quadrat is assigned or weighted at the time of sampling.

The sample variances are calculated as (Gillen and Smith 1986)

$$S_i^2 = \sum_{j=1}^{3} M_j^2 p_{ij}(1 - P_{ij}) - 2\sum_{j<k} M_j M_k p_{ij} p_{ik} \tag{6.12}$$

where S_i^2 = sample variance for ith species
P_{ij} = proportion of plots in which species i received rank j
P_{ik} = proportion of plots in which species i received rank k
M_j = multiplier of rank j
M_k = multiplier of species k

6.1.4 Productivity

An analysis of ecosystem processes, such as seasonal patterns of biomass dynamics, nutrient content, consumption, and decomposition, is dependent on an accurate estimate of net primary production. Above-ground plant material is usually considered as a measure of above-ground net primary production (ANP) and is estimated by harvesting standing crop when the system has apparently attained maximum standing crop. Productivity is the rate of biomass produced per unit time interval, while production is the amount of biomass accumulated over time. Standing crop, therefore, is weight of plant material present in a system at any given point in time. However, by the time measurements are taken, some material may have already senesced, detached, and disappeared from the observation point. The problem can be overcome by cumulative sampling. Cumulative sampling procedures utilize extensive seasonal sampling, and production is determined from a summation of maximum biomass estimates or positive changes in biomass. Lowest measured overwintering weight may be subtracted from the season's peak biomass to allow for carryover growth (Ovington et al. 1963; Singh and Yadava 1972, 1974). This approach to estimation of production increases field work but will increase accuracy of the estimate.

In tropical grasslands of India three distinct growth periods are distinguished (Singh 1968). In order to obtain an estimate of ANP within each growth period, standing crop values at the beginning of each growth period are subtracted from peak values. Above-ground net primary production is then the sum of these three seasonal production values.

Malone (1968) compared a peak standing crop method with the summation of species peaks method. A larger error term was associated with the summation of species peaks method, which ranged from 16 to 18% of estimates, compared with 7–11% for the peak standing crop method. Malone also found that differences between the two methods decreased as phenologies of dominant producers became more similar. Lauenroth (1970) found that the summation of species peaks in a mixed-grass prairie ecosystem gave an estimate of ANP that differed by only 7% from peak standing crop.

It is normally assumed that each species peaks only once in the growing season. However, it has been shown that certain species may have more than one peak during the growing season (Singh 1968). This problem can be overcome by summing positive increments in biomass by species or species groups. For example, Kennedy (1972) and Kelly et al. (1974) found that the summation of positive biomass increases gave maximum estimates of ANP. Kelly et al. also pointed out that estimates of ANP by summation of species peaks were lower when the sampling interval was increased from 1 to 4 weeks. This reduction in the biomass estimate is attributed to two factors: a reduction in the number of species detected and increased possibility of missing a positive biomass increment.

Most methods of estimating ANP ignore material that dies and becomes detached from plants before species peaks are observed. Therefore, the sum of positive increments in live biomass only is an underestimation because plant parts that senesced during sampling intervals are not accounted for. It way be argued that in a stable ecosystem, net biomass accumulation is zero during an annual period and, therefore, ANP should be equal to the annual amount of dead material disappearing. This assumption may not necessarily be the case for many ecosystems (Lester 1969).

Singh et al. (1975) compared several methods of calculating ANP. They found that most methods used in computing net aerial production were correlated with each other; methods that yielded highest or lowest estimates differed from site to site. Some methods appeared to be better discriminators of perturbations on vegetation, whereas others proved more useful for distinguishing sites or years. This indicates that no single method is universally applicable for estimating ANP.

Ground vegetation of forests contributes little to the total biomass. Yet, it is an important component of the ecosystem. The ground vegetation production (GVP) is calculated by one of the following two methods (Ford and Newbould 1977): (1) from seasonal trends of total GVP by subtracting minimum from maximum values or (2) from biomass trend of individual species when individual estimates are summed to calculate total GVP. Different calculations are used to determine the production of different species with the latter method. Species with perennial above-ground parts, production is calculated as maximum biomass minus initial biomass. Species with annual shoot production are taken as equivalent to maximum biomass. For species that are sparsely distributed and represented only occasionally through the season, production is considered as equivalent to maximum biomass recorded in any of the samples. The first method underestimates GVP when component species of ground vegetation do not develop maximum biomass at the same time of year. Ford and Newbould (1977) studied production of ground vegetation through a deciduous woodland cycle at five ages and found underestimates by the first method to be as follows: year 1, 55%; year 2, 24%; year 5, 30%; and year 9, 49%.

6.1.5 Considerations of Herbaceous Biomass Sampling

Plot Sizes and Shapes. An investigator frequently encounters the problem of choosing the best size and shape of a plot because such dimensions may influence estimates of biomass production. The main criteria for optimum plot size and shape are to obtain minimum variance and an unskewed data distribution. Plots should also be convenient to use in the field. In general, from a theoretical point of view, a much greater accuracy is achieved by making the plots as small as possible and by increasing the number of sampling units. There are, however, disadvantages in reducing the size of the plot below a certain limit. In large-scale survey work it would be prohibitive to select, demarcate, and collect data from a large number of plots. For small plots, greater portion of edge per unit area increases the difficulty of deciding which plants to include in the sample (Morris 1959, Sukhatme 1947). This bias is usually toward overestimation of the mean production value. The bias is more serious where plants are larger and fewer because incorrect inclusion or exclusion of a single individual influences total production to a greater extent than where plants are smaller and more numerous (Wiegert 1962).

Plots that are too small may also bias the observer's ability to identify the contribution of infrequent species unless a large number of observations are made. This bias occurs when dominant species occupy the entire plot. In contrast, plots that are too large result in excessive expenditure of time for a relatively small return of additional information. Increasing the size of plots reduces variance, and, thus, only a few plots would be included in the sample. This results in inflation of the confidence limits of the mean because of rapid increases in the "student's t value" for a few degrees of freedom. Thus, advantage of increasing the plot size to reduce variance is offset if the sampling scheme involves only a few plots.

The problem then is to find an optimum size of plot to sample a given vegetation type without having an adverse effect on accuracy as a result of an increase in size of the plot and reduction in the number of sampling units. Even when optimum size of the plot has been determined, a change in shape of the plot may affect sampling error. Methods to find the optimum size and shape are discussed later in this section. Except for research work, it may not always be feasible to spend time and money to find an optimum size and shape of a plot. Selection and use of a plot size and shape usually reflects the experience of an ecologist in a similar vegetation type.

Theoretically, long, narrow plots have smaller coefficients of variation when placed with their axis parallel to the gradient of heterogeneity (Christidis 1931). However, if direction of variability in soils or vegetation cannot be determined, then square or circular plots should be used.

Experimental errors remain constant or decrease as amount of biomass increases and results in smaller variances for larger plots. The coefficient of variation for bunchgrasses is normally much lower than that of sod-forming grasses (Van Dyne et al. 1963). If the material being sampled is fairly uniform, there will be little gain in accuracy by increasing the size of the plot.

Plots varying in size from 0.4 to 9.3 m² were tested by Campbell and Cassady (1949). After much trial, error, and analysis, they selected 0.9-m² circular plots. These 0.9 m² plots have been used extensively and provide satisfactory data on most grasslands. For sampling areas of uniform production, such as some improved pastures and old fields dominated by a few species, a plot that contains 0.10 m^2 may be satisfactory for practical use. However, Crafts (1939) found a long plot (0.6 × 7.5 m) to be more effective than the circle for estimating forage weight in southeastern United States rangelands.

In southern United States forested grasslands, a square plot of 1 m on a side is about the most convenient shape to use. On the other hand, circular plots are difficult to delineate in tallgrass types. A square or rectangle with one side removed is easy to use in all types of vegetation, particularly in tall brush or tallgrass types (Hughes 1959). In the western United States and for low-growing plants, circular or long and narrow plots are generally satisfactory for biomass measurements.

A 0.2-m^2 circular plot was considered most suitable on a foothills bunchgrass range in Montana, both statistically and timewise, for estimating herbage production (Van Dyne et al. 1963). Different sizes and shapes of plots used in the study and their relative efficiency percentages are given in Table 6.4. In the same study, rectangular plots of 0.4 m^2 (30 × 122 cm) and 0.6 m^2 (30 × 183 cm) were considered minimum sizes for estimating herbage production of shrubs and half shrubs, respectively. Characteristics of herbage weight data obtained from various sizes and shapes of plots on a bunchgrass range in Montana are given in Table 6.5. Plots that give the

TABLE 6.4   Relative Efficiency Percentages of Different Sizes and Shapes of Quadrats for Estimating Total Herbage Weight

Shape	Dimensions (cm)	Area (m^2)	Abbreviation[a]	Statistical Efficiency	Time Efficiency
		Red Bluff Ranch			
Circular	r[b] = 24	0.2	(C2)	100	100
Rectangular	30 × 60	0.2	(R2)	61	62
Square	43 × 43	0.2	S2	53	54
Circular	r = 17	0.1	C1	42	86
Long	5 × 183	0.1	L1	34	69
Rectangular	20 × 46	0.1	R1	18	36
Square	30 × 30	0.1	S1	7	1
		Shaw Ranch			
Rectangular	30 × 183	0.6	R6	100	71
Rectangular	30 × 122	0.4	R4	90	95
Rectangular	30 × 60	0.2	R2	48	100

From "Range herbage production estimates" by G. M. Van Dyne et al., *Ecology*, 1963, **44,** 748–753. Copyright © 1963 by The Ecological Society of America. Reprinted by permission.
[a] Used in Table 6.5.
[b] r = radius of the circle.

smallest estimates of the variance are the best to use from a statistical viewpoint. Then, for total herbage, the best plot is the circle (C2) listed in Table 6.4 as being 100% efficient.

Ahmed et al. (1983) used a 0.18 m² circular plot to estimate biomass in shortgrass vegetation of Colorado. A circle and a rectangle with areas of 0.18, 0.25, 0.32, and 0.50 m² were compared, and no significant differences were found between sizes or shapes in the estimate of biomass by clipping. The smallest area, therefore, should be chosen for use in measurement of biomass if there are no statistical differences among plot sizes.

TABLE 6.5 Characteristics of Herbage Weight Data Using Different Shapes and Sizes of Plots

	Plot[a]	Herbage Weight (g/m²) Mean	Variance	To Sample Herbage Weight Within 10% of Grand Mean Number of Plots	m² to Clip
		Red Bluff Ranch			
Total herbage	R1	87	3226	115	11
	R2	90	953	34	7
	S1	111	8815	313	31
	S2	82	1112	39	16
	C1	95	1387	49	5
	C2	79	585	21	8
	L1	91	1732	61	6
		Shaw Ranch			
Total herbage	R2	73	902	53	21
	R4	68	479	28	45
	R6	68	430	25	90
Grasses and sedges	R2	51	784	88	35
	R4	49	116	48	77
	R6	49	357	40	144
Forbs	R2	11	166	416	166
	R4	10	74	185	296
	R6	11	100	248	893
Shrubs and half-shrubs	R2	10	112	399	160
	R4	9	59	210	336
	R6	8	34	119	428

From "Range herbage production estimates" by G. M. Van Dyne et al., *Ecology*, 1963, **44,** 748–753. Copyright © 1963 by The Ecological Society of America. Reprinted by permission.
[a]Abbreviations same as defined in Table 6.4.

TABLE 6.6   Different Sizes and Shapes of Quadrats Used to Find Optimum Sizes and Shapes in a Study on a Bunchgrass Range in Northern Greece

Quadrat Size		Shape		
Number	Area (m²)	Square	Rectangular	Circular (Diameter)
1	0.0625	0.250 × 0.250	0.500 × 0.125	0.282
2	0.1250	0.354 × 0.354	0.500 × 0.250	0.399
3	0.2500	0.500 × 0.500	0.250 × 1.000	0.554
4	0.5000	0.707 × 0.707	0.500 × 1.000	0.798
5	1.0000	1.000 × 1.000	0.500 × 2.000	1.128

From Papanastasis. Optimum size and shape of quadrat. *J. Range Manage.* **30**: 447.

On a bunchgrass range of northern Greece, Papanastasis (1977) used five sizes of quadrats in three shapes to determine optimum size and shape of quadrat for herbage sampling. He found 0.0625- and 0.25-m² square plots to be more suitable for sampling herbage biomass on ungrazed and grazed ranges, respectively (Tables 6.6 and 6.7). Wiegert (1962) determined that 0.187- and 0.047-m² plots were optimum for sampling total green material and grasses, respectively, on old field vegetation in Michigan (Table 6.8). Note that grasses always had smaller variances than the other categories and that magnitude of variances did not change order for categories as plots increased in area.

Pechanec and Stewart (1940) found that rectangular plots were more efficient than circular plots on a native sagebrush–grassland in Idaho and considered a 12-m² (1.6 × 7.6-m) plot the most suitable sampling unit for estimating total herbage production. Optimum plot size was determined by Wassom and Kalton (1953) in Iowa. Biomass harvested in basic units of 1.1 × 1.2 m indicated that efficiency generally decreased as size of the plot

TABLE 6.7   Efficiency of Different Sizes of Plots Used to Find Optimum Size and Shape of Plots in a Study on a Bunchgrass Range in Northern Greece

Quadrat Size[a]	$\overline{X}$(g/m²)	S^2	Statistical Efficiency[b]	Time Mean ± Standard Error (min)	Time Efficiency[c]
1	323.34	47452.78	0.21	2.44 ± 0.11	1.00
2	363.20	72574.44	0.14	3.69 ± 0.16	0.66
3	366.74	25905.92	0.38	5.78 ± 0.20	0.42
4	350.44	31313.17	0.32	10.80 ± 0.43	0.22
5	311.85	9851.85	1.00	16.62 ± 0.48	0.15

From Papanastasis. 1977. Optimum size and shape of quadrat. *J. Range Manage.* **30**: 448.
[a]Quadrat sizes are the same as described in Table 6.6.
[b]Smallest variance divided by variance of the plot.
[c]Minimum time divided by time required to clip the plot.

TABLE 6.8 Weighted Variance of Mean and Weighted Mean for Four Categories of Vegetation and Five Quadrat Sizes[a]

Quadrat Size x(m^2)	Vegetation Category	Weighted Variance of the Mean V_m	Weighted Mean m(g)
1 (0.016)	Grass	0.968	4.68
	Forbs	2.079	5.99
	Total green	2.654	10.61
	Dead	21.535	45.79
3 (0.047)	Grass	0.243	3.82
	Forbs	1.895	4.42
	Total green	2.079	8.25
	Dead	20.707	41.09
4 (0.063)	Grass	0.318	3.92
	Forbs	1.743	4.82
	Total green	1.908	8.85
	Dead	16.738	36.04
12 (0.187)	Grass	0.139	3.60
	Forbs	0.578	3.54
	Total green	0.748	7.13
	Dead	11.473	37.70
16 (0.250)	Grass	0.150	3.72
	Forbs	0.785	3.89
	Total green	0.944	7.57
	Dead	11.656	38.32

From "The selection of an optimum quadrat size for sampling the standing crop of grasses and forbs" by R. G. Weigert, *Ecology*, 1962, **43**, 127. Copyright © 1962 by The Ecological Society of America. Reprinted by permission.

[a]All are calculated on the basis of a quadrat size of 0.25 m^2.

was increased. The basic unit of 1.1 × 1.2 m was found to be the most efficient.

Determination of Minimal Sample Area. The minimal sample area can be determined from a system of nested plots (Fig. 6.1). Initially, a small area, say, 10 × 10 cm (0.01 m^2), is laid out and total biomass is clipped. Then the sample area is progressively enlarged to twice the size, to 4 times, to 8 times, and so on. The additional biomass and time required to clip it is recorded separately for each enlarged area. The biomass, fixed cost (walking between stations, weighing, etc.), cost to clip each subplot in terms of time spent on clipping, and variance changes are recorded. The underlying principle used to estimate optimum plot size is to minimize total cost for a given variance, or alternatively minimize variance for a given total cost.

A method to determine optimum size of a plot that is independent of changes in the radius r was described by Wiegert (1962). The data on variances and costs associated with plots of various sizes were used to

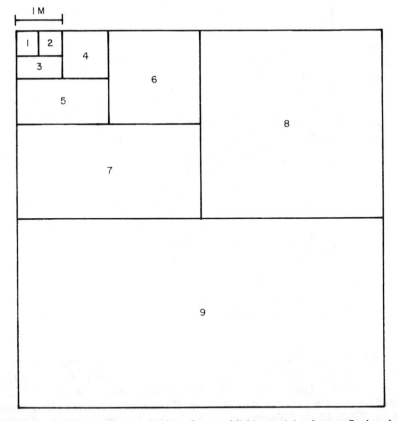

Figure 6.1 A system of nested plots for establishing minimal area. Each subplot numbered consecutively includes the area of the previous subplot. Thus, uneven-numbered subplots are square, and even-numbered ones are rectangular (*Aims and Method of Vegetation Ecology*, Mueller-Dombois and Ellenberg, Copyright © 1974 by John Wiley & Sons, Inc. Redrawn by permission of Wiley.).

calculate that plot size that provided minimum cost × variance product. The change in variance with change in plot size depends on the pattern of distribution of biomass production. Therefore, the dispersion or spatial distribution exhibited by vegetation biomass is important in selection of optimum plot size. This dispersion characteristic can best be determined by sets of nested quadrats that are randomly placed within the vegetation type.

The possible number of quadrats, used by Wiegert (1962), by addition of units was seven, but only five sizes were used for data analysis (Table 6.8). The relative quadrat size x equal to 1, 3, 12, and so on, describes the number of units of size 1 that are contained in the area of a given quadrat size. Means and variances of each stratum, for each category, can be used

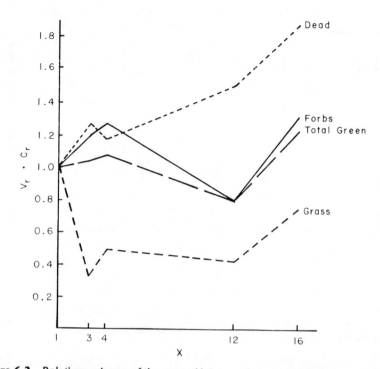

Figure 6.2 Relative variance of the mean V_r times relative cost C_r against the relative quadrat size x. Areas of the quadrat sizes are as follows: 1 (0.016 m²), 3 (0.047 m²), 4 (0.063 m²), 12 (0.188 m²), and 16 (0.250 m²). [From "The selection of an optimum quadrat size for sampling the standing crop of grasses and forbs" by R. G. Weigert, *Ecology*, 1962, **43**, 127. Copyright © 1962 by The Ecological Society of America. Reprinted by permission.]

to calculate a weighted mean m and a weighted variance of the mean V_m for each quadrat size. A relative weighted variance of mean V_r can be computed by dividing V_m of each quadrat size by the value of V_m of the smallest quadrat size. Data of relative variance of the mean V_r times relative cost C_r (see below), and relative quadrat size x are plotted to determine optimum sizes of quadrats for each category (Fig. 6.2) by selecting the lowest point or range. Optimum quadrat sizes were 0.187 m² for forbs and total green material, 0.047 m² for grass, and 0.016 m² for dead material. The rise in V_r for all categories between quadrat sizes 12 and 16 was attributed to a clumped distribution of plants (Evans 1952, Thompson 1958). It is assumed that a positive correlation existed between plant biomass on adjacent areas.

If the total cost of sampling equals C, then the cost of a single plot equals C/n, where n is the number of sample observations (plots). The cost C/n consists of a fixed cost, C_f (which is independent of the size of the plot and again, as previously noted, involves walking between stations, weighing, etc.) plus X times a cost C_v that represents time spent on collection of data from a plot. The relative cost C_r equals cost for a given plot size relative to cost for the smallest plot size. Then

$$C_r = \frac{C_f + XC_v}{C_f + C_v} \tag{6.13}$$

Values of cost factors determined in the study by Wiegert (1962) are given in Table 6.9.

Optimum plot size and shape were determined by Torrie et al. (1963). Estimates were computed to determine the number of replicates needed to detect differences of a given percentage among vegetation types of the mean at a given level of significance with assurance based on the relationship

$$r > |[2(t_0 + t_1)(CV)^2]|d^2 \tag{6.14}$$

where r is number of replicates per vegetation type, CV is the estimate of coefficient of variation, d is the difference to be detected expressed as percentage of the mean, t_0 is the t value associated with the level of significance, and t_1 is the t value corresponding to $2(1 - P)$, where P is the probability of detecting a difference. Variance of the mean $s_{\bar{y}}^2$ is determined by

$$s_{\bar{y}}^2 = \frac{s^2}{r(n)^b} \tag{6.15}$$

where s^2 is the variance among plots of one-unit size, r is the number of replicates, n is the number of basic units (plots to be added to obtain a larger plot) under consideration, and b is the coefficient of heterogeneity

TABLE 6.9 Cost Data (min) for the Various Quadrat Sizes

Cost Component	Quadrat Size X[a]				
	1	3	4	12	16
Fixed cost C_f	10	10	10	10	10
Variable cost $C_v X$	2	6	8	24	32
Total cost of the quadrat	12	16	18	34	42
Relative cost C_r	1.00	1.33	1.50	2.83	3.50

From "The selection of an optimum quadrat size for sampling the standing crop of grasses and forbs" by R. G. Weigert, *Ecology*, 1962, **43**, 127. Copyright © 1962 by The Ecological Society of America. Reprinted by permission.

[a]Quadrat sizes are the same as described in Table 6.8.

(Smith 1938). Optimum plot size for minimization of variances in Smith's study was

$$H_{opt} = \frac{bC_1}{(1 - b)C_2} \qquad (6.16)$$

where C_1 is a fixed cost per plot regardless of size and is proportional to the number of plots per treatment and C_2 is the cost dependent on plot size and is proportional to the total area per vegetation type. Areas harvested consist of adjacent strips of any dimension. Each strip is divided into a number of smaller units so that the smallest unit may be 0.1×1.0 m. These units are combined in various ways to determine the effect of plot shape. As plot size increases, variance among plots decreases and variances among sampling units within plots increases. These changes are more pronounced as plot size increases lengthwise rather than in width. Smith found the optimum plot size was 1.26 basic units when $C_1 = 70\%$ and $b = 0.35$. Analysis of coefficients of variability indicated that optimum plot size was 1.4×3 m.

A method for the estimation of optimum plot size from experimental data, using theoretical considerations, was derived by Koch and Rigney (1951). Plots of successively larger sizes are formed by combining adjacent units. See Fig. 6.1. This increases the variance of plot means and makes

$$s_{\bar{y}}^2 = \frac{s^2}{(n)^b} \qquad (6.17)$$

where b is an index of soil or other environmental heterogeneity ranging from 0 to 1, with larger values indicating lower correlations. In logarithmic form this is

$$\log s_{\bar{y}}^2 = \log s^2 - b \log n \qquad (6.18)$$

If K_1 = proportion of total cost dependent on number of plots per vegetation type and K_2 = proportion of total cost dependent on total area per treatment, then the optimum plot size has n units, where n is given by

$$n = \frac{bK_1}{(1 - b)K_2} \qquad (6.19)$$

Variances for different plot sizes are obtained from components of variance analysis for plots, the size of blocks (blocks mean square), and plots within the entire area (error mean square). A least-squares estimate of b is obtained and then can be used to determine the number of n units.

6.2 SHRUB BIOMASS

Shrubs constitute an important part of many ecosystems and provide browse, not only for wild animals but also for various categories of domestic livestock. In fact, browse may constitute the bulk of available forage in some ecosystems. In other ecosystems, it may be the only available forage during certain times of the year.

An accurate assessment of shrub biomass is also important for an evaluation of the productivity of an ecosystem, cycling of nutrients and energy. An estimate of shrub biomass is also required to describe the amount of fuel for firewood, and appraisal of flammability. To evaluate shrub flammability, it is necessary to know not only shrub biomass weight but also its size and distribution over an area. The weight of foliage is also useful for an estimation of transpiration use of water by shrubs.

Browse measurement is one of the more difficult components of vegetation to determine. Similar to the measurement of herbaceous production, many techniques have been used to measure quantities of browse. All are based on one of three standard procedures: direct harvest method, indirect weight estimates, or a combination of harvest and estimates. Clipping and weighing is generally the most accurate method but is also laborious. A large number of plots are often necessary to measure shrub production. Indirect methods for reliable estimates may involve less time than direct harvesting and permit a large number of observations to be obtained at a relatively low cost.

6.2.1 Harvest of Current Annual Growth

This technique involves clipping and weighing current annual growth of twigs and leaves. The results are usually expressed as oven-dry weights. Blair (1959) measured browse production by harvesting five 0.3-m^2 plots within 0.04-ha replicates. He found this approach to be inadequate for estimation of mean production of total browse within a sampling error of ±20% at a 90% level of probability.

Bobek and Dzieciolowski (1972) studied the methods of browse biomass estimation in different types of forests in central Europe and also found that accuracy of the estimates was low due to high variability of browse distribution in an area. Then it is evident that a large number of clipped plots is usually necessary for measurement of browse biomass production with an acceptable level of reliability.

6.2.2 Nondestructive Methods of Shrub Biomass

Limitations of direct harvest techniques found in obtaining efficient estimates of shrub production led to development of indirect or nondestructive methods for shrub production. Measurements of crown area, circumference, diameter, and volume; plant height; basal stem diameter; twig diameter, length, and weight; and width of xylem rings, alone or in combination with other variables, have been used to estimate shrub production. The indirect measurement techniques can be divided into two broad categories: reference unit technique and dimension analysis. The reference unit technique involves matching a known unit of weight against observations. In the dimension analysis technique, a mathematical relationship is developed between easily obtained plant dimensions and foliage weight. Specific plant dimensions useful in estimating biomass of shrubs are discussed individually.

Reference Unit Technique. A small unit of plant such as a shoot of a given dimension is designated as reference unit (Fig. 6.3). The size of the reference unit should be 10–20% of the foliage weight of the average plant (Andrew et al. 1979, Kirmse and Norton 1985). A few reference units are clipped, and average green or dry green weight is determined. The next step is to count or estimate similar reference units. The number of estimated reference units is multiplied by the average weight of clipped reference units to estimate shrub biomass production. The actual sampling is preceded by a short period of training. During training the technicians familiarize themselves with a reference unit and the number of such reference units on different plants. Estimates are frequently checked with actual weights. The reference unit technique is not suitable for shrubs which have compact, dense, unsegmented growth form (Fig. 6.4). The technique was tested by Andrew et al. (1981) on two small shrubs in Australia and compared favorably with other techniques for estimating shrub biomass.

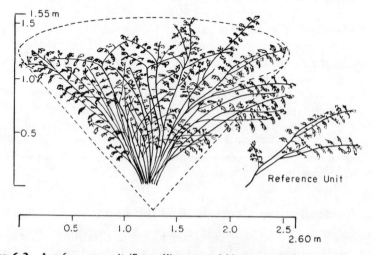

Figure 6.3 A reference unit (From Kirmse and Norton. 1985 "Comparison of reference unit method and dimension analysis methods." J. Range Manage. 38: 426.).

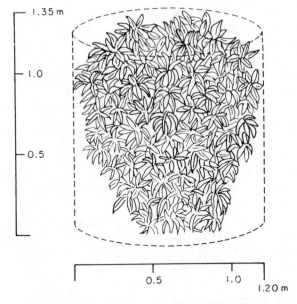

Figure 6.4 Example of shrub growth not suitable for reference unit method for estimation of biomass (From Kirmse and Norton. 1985 "Comparison of reference unit method and dimension analysis methods." J. Range Manage. 38: 426.).

Kirmse and Norton (1985) compared the reference unit method to dimension analysis methods for two large shrubs in Australia. The amount of biomass variation accounted for by the reference unit approach ranged from 89 to 98 percent. The R^2 values obtained in applying the dimension analysis method were .937 and .948. The two methods took about the same amount of time. Therefore, Kirmse and Norton recommended the reference unit method for field application, rather than dimension analysis.

Crown Area. Some studies have shown high correlation between the dimension of crown area and leaf biomass. For example, Goebel et al. (1958) found that cover values accounted for 90% or more of the variation of production of shrubs growing in pure stands in the western United States. Correlation of shrub biomass production with growth and ground cover was studied in Nevada by Kinsinger and Strickler (1961). Crown area, length of current year's twig elongation, and production were measured for several consecutive years in plots 5 m² in area (0.6 × 7.6 m). Production was determined by clipping varying percentages: 25, 50, and 75% of current annual growth was removed in spring; 10–30% in summer; and 40, 60, and 90% in winter. Then 20–50 random measurements of current growth were made. Crown cover measurements, using the area-list method of Pearse (1935), were made with an area ruler prior to clipping. Correlation and regression coefficients were calculated between twig growth and production, growth and crown cover and production for each year, and location and season of clipping. Highly significant correlations were found between spring twig growth and production. Therefore, an estimate of production from the current year's twig length would be more accurate from measurements made in spring. Crown cover X was significantly correlated with production $Y(r = 0.4, Y = 40.0 + 9.8X)$ and indicated that winter was the best time to estimate yield from shrub ground cover. Percent crown cover was recorded for three seasons on 5-m^2 plots replicated 3 times at three locations in Nevada. Ground cover measurements were made with an area ruler prior to clipping. Growth was measured as length of the current year's twig elongation. Highly significant correlations were found between spring growth and production ($r = .804$).

Medin (1960) used crown diameter to estimate shrub production on different sites in northwestern Colorado. Crown diameter of all plants was measured on each 40-m^2 plot. One plant per plot was randomly selected and clipped to obtain air-dry weights for all current annual twig growth after cessation of annual growth. The regression of production with crown diameter yielded a correlation coefficient of $r = .84$ (log weight = 1.25 + 1.66 log of crown diameter). Harniss and Murray (1976) used crown circumference C times height H to develop the mathematical model

$$DW = 0.0167HC^{1.25} \tag{6.20}$$

to predict the dry-leaf weight DW of shrubs. Figure 6.5 shows the predictor in three-dimensional form. The model was tested at two different sites and it accounted for 93 and 80% variation, respectively. The results suggest that the geometric shape of the predictor is representative of height–circumference/weight relationship in big sagebrush and that the predictor may be used for other stands.

Murray and Jacobson (1982) used dimension analysis for predicting biomass of several shrubs. A simple linear model of the following form gave the best biomass predictions:

$$\ln Y = \ln a + b \ln H + c \ln C \tag{6.21}$$

where Y is the weight (g); H is the height (cm); C is the circumference (cm); and a, b, and c are the Y intercept and slope coefficients, respectively. This model worked well in prediction of not only the leaf biomass but also live and dead twig biomass components. Note that equation 6.21 is the log transform of equation 6.20.

Crown Volume. Crown volume has been shown in many studies to be an adequate predictor of the total leaf biomass of shrubs. At least two mea-

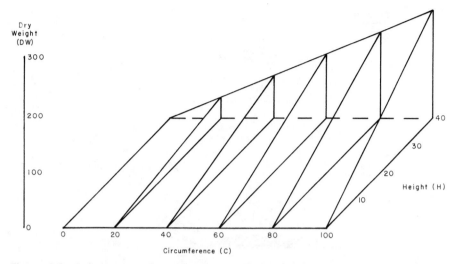

Figure 6.5   Relationship of height and circumference to dry leaf weight of sagebrush ($DW = 0.0167HC^{1.25}$) (From Harniss and Murray. 1976 "Reducing bias in dry leaf weight estimates of big sagebrush." J. Range Manage. 29: 431.).

surements (diameter and height) of crown are required in addition to foliage weight to develop volume–weight relationships. In the case of shrubs with irregular outline, an average of two diameter measurements are required to calculate volume. One diameter is recorded in the longest direction of the canopy and the other in the perpendicular direction to the first measurement. The calculations of volume depends on the geometric configuration of the shrubs. That is, shapes of conical, cylindrical, spherical, and so forth are considered. Some canopy shapes and geometric formulas to fit these shapes are given in Figure 6.6.

The crown volume–weight relationships are normally developed for individual species and are applicable to areas in which these are developed. However, Bentley et al. (1970) attempted to develop crown volume–weight relationships applicable to several species. An important consideration in computing crown volume is the openings in canopy. Canopy openings less than 30 cm were considered as continuous intercept by Rittenhouse and Sneva (1977). They developed mathematical models to predict the leaf and woody biomass of shrubs from height and two measurements of crown width. The R^2 values for different models ranged from 0.72 to 0.95.

In western Montana, Lyon (1968) estimated shrub twig production from crown volumes on eleven different sites. Shrub crowns were measured in the summer while plants were in full leaf. One measurement was made through the long dimension of the canopy and another at right angles (diameters a and b, respectively). Height h was measured from ground

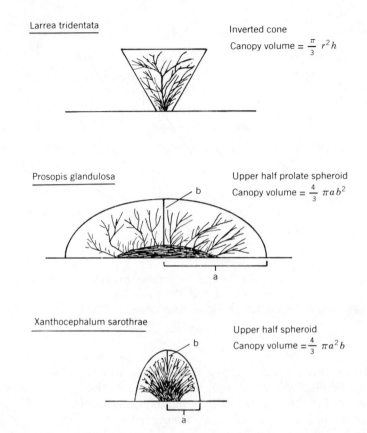

Larrea tridentata

Inverted cone
Canopy volume $= \frac{\pi}{3} r^2 h$

Prosopis glandulosa

Upper half prolate spheroid
Canopy volume $= \frac{4}{3} \pi a b^2$

Xanthocephalum sarothrae

Upper half spheroid
Canopy volume $= \frac{4}{3} \pi a^2 b$

Figure 6.6 Canopy shapes and equations to describe shapes (Ludwig et al. 1975).

surface. Volume was calculated using the equation

$$V = \left(\frac{\pi}{4}\right) abh \tag{6.22}$$

Current annual growth on each plant was clipped after leaf fall to obtain an estimate of twig production. Total annual growth was collected for plants with volume less than 2 m³, while a random quarter of the shrub canopy was clipped for larger shrubs and the other three quarters were estimated. Results of regression analysis indicated that twig production on individual sites can be predicted accurately. Over 80% of the variation in twig production was associated with shrub volume for 6 out of the 11 sites. It was noted that prediction of biomass was least accurate where soil surface disturbance had occurred recently. Therefore, unnatural factors for a site should always be noted, and plants should be measured by strata when possible. That is, regression equations should be developed for each uniform environmental condition to obtain the best results.

In the shrub–steppe region of southeastern Washington, Uresk et al. (1977) used a double-sampling procedure to estimate biomass of different shrub components of sagebrush (*Artemisia tridentata*). Plant dimensions were related by regression analyses to harvest categories (leaves, wood, etc.). The following measurements were recorded for each plant: (1) longest diameter of the canopy, (2) longest diameter of the canopy measured at right angles to the previous dimension, and (3) maximum height. Shrubs were cut at ground level, and oven-dried weights were obtained for various biomass compartments separated by hand. The same data were visually estimated on a large number n' of randomly selected sagebrush plants. Data from clipped shrubs n were used to adjust the visual estimates through linear regression. A separate equation was developed for each harvest category.

One objective of the study was to minimize variance of estimated mean phytomass for each category for a fixed cost. It was assumed from experience that a clipped estimate of biomass was 120 times as expensive to obtain as were the dimension measurements. The optimum ratio of n' to n was estimated using the technique described by Cochran (1977), as in Equation (6.4) (above):

$$\frac{n}{\sqrt{V_n c_{n'}}} = \frac{n'}{\sqrt{V_{n'} c_n}}$$

where V_n and $V_{n'}$ are the clipped and guessed biomass weight variances, c_n and $c_{n'}$ are the costs associated with the respective sample sizes n and n'. The objective is to find an optimum number of n to n' measurements.

Generally, only one variable at a time, usually volume, is used in the regression to estimate the biomass of interest. The double sampling ds estimate is obtained by Equation (6.1), which had different notation

$$\overline{Y}_{ds} = \overline{Y}_n + b(\overline{X}_{n'} - \overline{X}_n)$$

where $\overline{X}_n$ is the mean volume based on $N = 20$ clipped shrubs and $\overline{X}_{n'}$ is the mean volume per shrub of the $n' = 569$ shrubs in the eight plots used. The variance is estimated by

$$\text{Var}(\overline{Y}_{ds}) = S_{yx}^2 \left[\frac{1}{n} + \frac{(\overline{X}_{n'} - \overline{X}_n)^2}{\Sigma_{i=1}^{20}(\overline{X}_{i'} - \overline{X}_n)^2} \right] + \frac{S_y^2 - S_{yx}^2}{n'} \tag{6.23}$$

The average biomass of shrub per square meter ($\overline{\overline{Y}}$) is estimated by multiplying the average biomass per shrub $\overline{Y}_{ds}$ by average number of shrubs per square meter $\overline{Z}$.

$$\overline{\overline{Y}} \simeq \overline{Y}_{ds}\overline{Z} \tag{6.24}$$

Its variance is estimated by

$$\text{Var}(\overline{\overline{Y}}) \simeq \overline{Z}^2 \, \text{Var}(\overline{Y}_{ds}) + \overline{Y}_{ds}^2 \, \text{Var}(\overline{Z}) \tag{6.25}$$

The optimum ratio of n' to n can be obtained from the equation listed on p. 233 and rearranged to obtain

$$\frac{n'}{n} = \sqrt{\frac{(S_y^2 - S_{yx}^2)/C_{n'}}{S_{yx}^2/C_n}} \tag{6.26}$$

which is Equation (6.4) rearranged. The highest correlations were obtained with volume (length × width × height) and length, which were the variables chosen for the regressions.

A cost ratio of C_n to $C_{n'}$ is 120:1. That is, it was 120 times as expensive to clip as to measure dimensions, and the total cost was $1000. This resulted in optimum sample sizes of $n = 7$ and $n' = 184$. It was observed that different categories of plant parts had different optimum allocations. A summary of the results of regression analysis by Uresk et al. and estimated optimum ratios are given in Table 6.10.

Bryant and Kothmann (1979) studied the variability in predicting edible crown leaf browse on 12 different species in western Texas. Biomass esti-

TABLE 6.10 Volume Measurements and a Double-Sampling Technique for Estimation of Sagebrush Phytomass

Dependent Variable	R^{2a}	n	g/Shrub ±SE	g/m² ±SE	Estimated[b] Optimum Ratio $n':n$	Reduction in Variance (%) Under Optimum Allocation
Leaves	.68	20	65 ± 10	10 ± 2	16:1	58
Livewood	.80	20	272 ± 48	43 ± 10	23:1	73
Flowering stalks	.52	19	33 ± 19	5 ± 3	12:1	42
Deadwood	.80	19	49 ± 40	8 ± 7	10:1	33
Total phytomass	.86	20	440 ± 70	69 ± 16	28:1	80

From Uresk et al. 1977. Sampling big sagebrush for phytomass. *J. Range Manage.* **30**: 313.
[a]R^2 significant at $P < .01$.
[b]n = clipped shrubs and dimension measurements, n' = plants on which dimensions were measurements and phytomass visually estimated.

mates were made with regression techniques using crown volume–weight relationships. They concluded that a log–log function may yield the best results for large species. That is

$$\log \text{weight} = a + b \log (\text{crown volume}) \qquad (6.27)$$

where a and b are the usual linear regression model coefficients. Small species may require a quadratic function for best results.

The quadratic form is

$$\log \text{weight} = a + b(cv) + c(cv)^2 \qquad (6.28)$$

or

$$\log \text{weight} = a + b(cv)^2 \qquad (6.29)$$

where cv is crown volume and the coefficients are as defined previously. Species with irregular form and very little foliage may require other mathematical relationships.

Stem Diameter. The relationship of total above-ground weight and leaf weight to basal stem diameters was determined by Brown (1976) for 25 northern Rocky Mountain shrub species. Highest correlations were obtained for natural logarithm of basal diameter and total above-ground weight and leaf weight for all species. Addition of natural logarithm of stem length did not increase precision greatly, even though its contribution in accounting for biomass variation was significant. Regression components of selected species are given in Table 6.11. Note that the overall variation increases and that the log (basal diameter) does not account for as much of the weight variation (measured by R^2). This decrease occurs because the relationship between log weight and log stem basal diameter is not consistent for all species. Brown recommended that geometric means of diameter ranges be used for both low- and high-diameter classes, while medium-diameter classes could be converted to either an intermediate value or a geometric mean.

Twig Measurements. It is usually more convenient to predict browse production from twig length, which can be measured accurately and quickly, compared to some other methods. Branch dimensions were used by Whittaker (1965) to estimate branch production of woody species in the Great Smoky Mountains of the eastern United States. Shrub and tree biomass production were predicted by counting and measuring branches from stem apex down and data were recorded for a number of variables. A matrix of mean coefficients of correlation among different variables is given in Table 6.12 showing high correlations among many variables. However, the pre-

TABLE 6.11 Regression Components for Estimating Leaf and Total Above-Ground Weights of Sample Woody Shrub Species and Three Shrub Groups Using Linear Equation

$$\ln (\text{weight, g}) = a + b \ln (\text{basal diameter, cm})$$

Species	n	Range of Sample Diameters (cm)	Leaf Weight			Total Above-Ground Weight		
			a	b	r^2	a	b	r^2
Low Shrub								
Symphoricarpos albus (L.) Blake (snowberry)	31	0.2–1.2	1.848	1.721	.68	3.490	2.285	.88
Vaccinium globulare Rydb. (blue huckleberry)	44	0.3–1.7	1.480	2.537	.84	3.388	3.150	.97
Combined species (5)	295	0.2–1.7	2.033	2.165	.67	3.565	2.667	.91
Medium Shrub								
Artemisia tridentata Nutt. (big sagebrush)	22	0.8–6.9	1.603	1.888	.75	3.161	2.242	.92
Juniperus communis L. (common juniper)	23	0.8–2.9	3.414	1.650	.80	4.081	2.202	.92
Combined species (5)	226	0.3–6.9	1.945	2.363	.66	3.580	2.853	.92
High Shrub								
Amelanchier alnifolia Nutt. (serviceberry)	39	0.4–4.5	1.691	2.111	.83	3.607	2.887	.99
Acer glabrum Torr. (mountain maple)	31	0.4–3.7	1.868	2.038	.89	3.634	2.752	.98
Combined species (4)	222	0.4–6.3	1.930	1.974	.79	3.507	2.697	.95

From Brown. 1976. Estimating shrub biomass from basal stem diameters. *Can. J. For. Res.* **6:** 154.

diction of branch weight for evergreen and deciduous shrubs should use branch diameter and functions of branch diameters as indicated by coefficients listed in column 2 of Table 6.12. Correlations were found to be useful for the estimation of dry weights of branch wood with bark, current twigs and leaves, and branch production from other measurements. The highest correlation for branch weight (wood with bark) was with powers of the diameter. Current twig weight (with leaves) was strongly correlated with the number of twigs and with the branch dry-weight–age relation.

Ratios of twig length to twig weight were recorded by Halls and Harlow (1971) at 1-year intervals over a 5-year period in Texas and Virginia. Total length was the single best variable for predicting weight ($r = .977$). However, ratios of twig weight to length were inconsistent among years and locations. Basile and Hutchings (1966) determined twig diameter–length–weight relations of bitterbrush (*Purshia tridentata*) in Idaho. Regression

TABLE 6.12 Coefficients of Correlation for Branches of Shrubs, Great Smoky Mountains[a]

	1	2	3	4	5	6	7	8	9
Means	.50	.71	.74	.74	.70	.71	.73	.69	.70
1. Age		.58	.51	.61	.70	.62	.57	.52	.56
2. Branch weight	.52		.91	.82	.77	.72	.88	.90	.93
3. Weight/age	.46	.92		.80	.74	.70	.87	.89	.91
4. Diameter	.60	.82	.76		.81	.97	.97	.91	.88
5. Length	.61	.75	.72	.81		.80	.78	.72	.82
6. Log diameter	.60	.69	.66	.95	.82		.89	.80	.77
7. 4^2	.54	.88	.80	.96	.74	.84		.98	.95
8. 4^3	.47	.86	.78	.88	.64	.72	.98		.97
9. $4^2 \times 5$	.52	.93	.84	.88	.79	.73	.95	.94	
Means	.53	.70	.75	.74	.73	.72	.70	.64	.68

From "Branch dimensions and estimation of branch production" by R. H. Whittaker, *Ecology*, 1965, **46**, 366. Copyright © 1965 by The Ecological Society of America. Reprinted by permission.

[a]Upper-right diagonal matrix, mean coefficients for branch samples from evergreen shrubs; lower-left diagonal matrix, for branch samples from deciduous shrubs. Top and bottom lines are means of correlation for one variable with all other variables for evergreen shrubs at the top and deciduous at the bottom.

analyses were conducted for biomass weight W on stem diameter D, length L, and diameter + length ($D + L$). Twig weight was highly correlated with length ($r = .86$, where $W = -0.063 + 0.057L$) and with diameter + length ($r = .95$, where $W = -.22 + 4.56D + .0301L$). Provenza and Urness (1981) studied these relationships for blackbrush (*Coleogyne ramosissima*) in western Utah. In both cases, the correlation between twig diameter, length, and weights was sufficiently consistent to estimate browse production.

Ferguson and Marsden (1977) also used twig diameter, length, and weight relations to estimate browse production of bitterbrush. Thirty shrubs were randomly selected, and each shrub canopy was visually divided into lower and upper halves and into north and south halves, thus quartering the canopy. Three unbranched twigs taken over a wide variety of twig lengths were collected from each quarter. Length and diameter were measured. Mean diameter was calculated from two measurements taken perpendicular to each other. Regression analysis showed a high correlation ($r = .752$ to $.807$) of length to diameter and weight to diameter squared ($r = .893$ to $.929$).

Twig and Stem Measurements. Twig and stem measurements for estimating browse were used by Schuster (1965) for eight species in southern Texas. Shoot weights (twigs and leaves) and twigs (without leaves) represented summer and winter browse, respectively. Measurements recorded were main stem diameter at groundline, number of live twigs per plant, total length of twigs, and total weight of twigs with and without leaves. The best prediction of biomass production ($r = .968$) was a combination of twig numbers and lengths. The relationship was, however, not consistent among species. Twig length was the single variable most closely correlated with both shoot and twig weight for most species. Twig counts had the lowest correlations with weight. Stem diameter squared was also highly correlated to weight.

The twig count method to measure biomass of hardwood browse for deer was used by Shafer (1963) in Pennsylvania. Shafer compared the twig count method with weight estimates and the clip-and-weigh method. Thirty circular plots, 9.2 m^2, were set up to include all browse between ground level and 1.8 m. Ocular estimates of green weight of browse, total

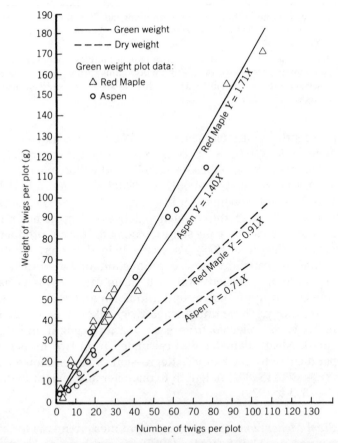

Figure 6.7 Results of the clip-and-weigh method, showing the relationship of weight of twigs per plot to number of twigs per plot for red maple, and red oak (Redrawn from E. L. Shafer, Jr. "The twig-count method for measuring hardwood deer browse," 1963, *J. Wildl. Manage.*, 27: 428–437. Copyright © 1963 by The Wildlife Society.).

and by species, within a plot were made and browse was clipped. Results of the clip-and-weigh method are shown in Figure 6.7. There were no significant differences among the three methods. There were differences, however, in the average time required per plot for each method. Results indicated that the twig count method was almost as rapid as the weight estimate and just as accurate as the clip-and-weigh method. A summary of the data analysis is given in Table 6.13.

Bartolome and Kosco (1982) described an "architectural model" for estimation of browse productivity of deerbrush (*Ceanothus integerrimus*) and other shrubs. Their model describes the stem and branch architecture in terms of first-, second-, third- and fourth-order branching. The primary stem that forms the main supporting structure is first order. The lateral branches arising from the main stem is second order, and so on. Basal diameter of the branches (second order) arising from the primary stem predicted leaf and branch weights with $r^2 = .97$ using an allometric transformation in a linear regression. The model is of the form

$$y = ax^b \tag{6.30}$$

or

$$\ln y = a + b \ln X \tag{6.31}$$

TABLE 6.13 Number of Plots and Time Required for Estimating Biomass by Three Methods at the 95% Probability Level for Species in Figure 6.8

Sampling Method	Number of 100-ft² Circular Plots	Total Plot Time (hr)
5% Accuracy		
Weight-estimation	648	34
Twig count	639	49
Clip and weigh	596	216
10% Accuracy		
Weight-estimation	162	8
Twig count	159	12
Clip and weigh	149	54
20% Accuracy		
Weight estimation	41	2
Twig count	39	3
Clip and weigh	37	13

From E. L. Shafer, Jr. 1963. The twig-count method for measuring hardwood deer browse. *J. Wildl. Manage.* **27**: 435. Copyright © 1963 by The Wildlife Society.

where y and x are branch weight and diameter, respectively, within a given order of branching.

Growth Rings. Width of growth rings has been used in some cases to determine average annual productivity of shrubs. Ring width differences within a site normally do not vary much, but site-to-site differences may be great. Some woody plants, however, do not have well-defined concentric growth rings. Davis et al. (1972) reported 20% difference within site and as much as 600% among sites for width of rings. Younger plants have larger growth rings as compared to older plants. This may result in an overestimation of production. To eliminate the problem caused by younger plants with larger rings, individual ring widths are standardized by dividing each ring width by average ring width of that particular plant. The growth ring widths are then expressed as a percentage of the average width. Production of a site should then be expressed as a percentage of the site's average annual production. The stem or branch cross sections may be stained with iodic green or sanded and polished to measure growth ring widths.

Xylem rings usually increase at three heights of the plant: base of the shoot, 2 cm above surface of forest litter, and in the middle and upper 2-year-old part of the shoot as measured by Moszynska (1970). Production can be estimated by multiplying the ratio of area of the current year's xylem ring to the area of cross section of the shoot by the biomass of old shoots:

$$W = \frac{D^2 - d^2}{D^2} \qquad (6.32)$$

where D is the diameter of the shoot and d is the diameter corresponding to growth in previous years. Davis et al. (1972) also estimated browse production from shrub ring widths in Nevada. Productivity was determined by clipping current year's growth on twenty 0.88-m^2 plots. Stems were measured to the nearest 0.01 mm with a Craighead–Douglas dendrochronograph. Ring widths were measured along two radii on each cross section and averaged for each year. Plants measured were winterfat, shadscale, bud sagebrush, spiney hopsage, and big sagebrush. The factors were analyzed by all possible correlations followed by stepwise regression with production as the dependent variable. Many of the factors correlated significantly with production, but widths of growth rings of shrubs accounted for most of the variation in production. Correlation coefficients r of production with the width of growth rings ranged from .615 for bud sagebrush to .972 for big sagebrush (Table 6.14). Big sagebrush and shadscale ring widths varied exponentially with production, while a linear relationship expressed production from ring widths for the other shrubs. The linear regression probably represented only a portion of the complete

TABLE 6.14 Production Estimation Y Equations with Ring Widths X of Shrubs as the Independent Variable

Shrubs	Equation	r^{2a}
Winterfat (*Cerotoides lanata*)	$Y = 1620X - 129.22$	.882
Shadscale (*Atriplex confertifolia*)	$\log Y = \log(X)1.454 + 3.45$ $Y = 2795X^{1.454}$	.870
Bud sagebrush (*Artemisia*)	$Y = 756.43 + 39.33X$	.378
Spiny hopsage (*Grayis spinosa*)	$Y = 1328.69 - 76.41X$	.693
Big sagebrush (*Artemisia tridentata*)	$\log Y = \log(X)1.5196 + 2.54$ $Y = 346X^{1.5196}$	.678
Big sagebrush	$\log Y = \log(X)1.55 + 2.55$ $Y = 354X^{1.55}$	.945

From Davis et al. 1972. Estimating forage production. *J. Range Manage.* **25:** 400.
[a]All *r* values significant at $P \leq .01$.

curve. Figure 6.8 illustrates growth rings for two big sagebrush plants. Growth ring widths predicted that the plant in Figure 6.8*a* current year's growth was 240 kg/ha, while the actual production was 218 kg/ha. Figure 6.8*b* illustrates a plant's growth rings that predicted 100 kg/ha compared to an actual of 106 kg/ha. Predictions were within 10% and 6% of actual production, respectively.

6.2.3 Considerations of Shrub Biomass Sampling

Identifying Current Growth. An important concern in the measurement of browse production is the definition of what plant parts constitute usable and available forage. Therefore, many field technicians tend to develop their own methods of measurement standards that result in a lack of a common basis for data. It is difficult to objectively define usable and available browse because animal species preference, season, and grazing pressure all determine amount of utilization. Availability of forage is influenced by terrain, season, density of vegetation, cover, height of browse from the ground, and growth form of plants. Brown (1954) suggested that production of shrubs should be based on that portion of the current year's twig growth that is edible and available to animals. Quite often, bark of new growth is either different in color from that of the old growth or covered by hairs. In contrast, color of old growth appears grayish in color and the epidermal texture is different when compared to new growth. In some species, leaves occur on the current year's twigs only. However, this is not

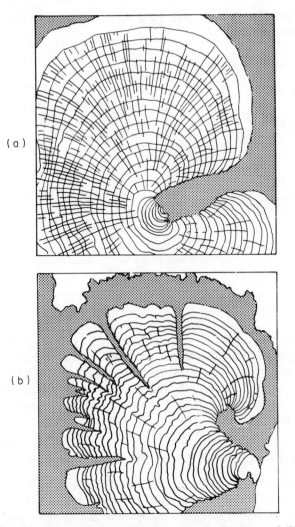

Figure 6.8 Sagebrush growth rings: (a) average growth ring = 0.73 mm; (b) average growth ring = 0.41 mm (From Davis et al. 1972 "Estimating forage production." J. Range Manage. 25: 401.). (See text for estimated and actual production.)

true in the case of evergreen species. A ring of bud-scale scars on some species or a slight swelling of the stem on other species may denote the point of origin of current growth. In actual practice, the ability to distinguish accurately between old and new plant growth is largely a matter of experience and familiarity with growth characteristics of the species.

On deer ranges in southern Michigan, Westell (1954) included twigs of 6 mm or less in diameter in available yield under normal deer use, and twigs up to 1.25 cm in diameter were included in production under heavy deer population pressure. Leaves and twigs represented summer production, while twigs alone represented winter availability. Gysel (1957) also used twig diameter to estimate forage biomass but confined measurements of current growth to twigs 3 mm or less in diameter. Forage up to 1.8 m above ground was considered available to deer. In the Missouri Ozarks, Dalkes (1941) confined sampling of current annual growth of twigs up to 5 mm in diameter and measured production up to a height of 1.5 m. Lay (1956), in eastern Texas, also confined measurements up to 5 mm but increased height of available forage to 1.8 m. Campbell and Cassady (1955) recommended that shrub production be measured up to 1.4 m above ground on southern livestock and deer ranges in the United States. For the same ranges, Blair (1959) suggested a height of 1.5 m from the ground. In a mixed oak–pine stand in Virginia, Barrett and Guthrie (1969) considered browse to be current annual terminal growth of all woody species from the ground to a height of 1.5 m. A standard range survey procedure has been to consider current annual twigs within 0.8–1.5 m of height aboveground level available as forage. However, densely clumped growth of woody plants, and dense short twigs induced by close browsing, pose problems in the determination of forage available.

Wight (1967) studied the effect of plot size on yield estimates of Nuttal saltbrush (*Atriplex nuttallii*) yields in Wyoming. A 18 × 18-m plot of Nuttal saltbrush was completely harvested using 0.3 × 0.3-m sampling units. Contiguous 0.09-m^2 plots were combined into larger units of various sizes and shapes, and yield data were examined. Wight found that the size of the sampling unit had a pronounced effect on estimated yield variation. Increasing the sampling unit from 0.3 to 5.5 m^2 caused the coefficient of variation ($100s/\bar{x}$) to decrease from 134 to about 30%, while enlargement beyond the 5.5-m^2 size had little effect on the coefficient of variation. The degree of aggregation of the saltbrush was responsible for the lower variation with the larger plot size.

Efficiencies in estimating shrub yields can be gained by using long narrow plots when shrub individuals are clumped or aggregated. Then, more variability in biomass is included within each plot compared to that found in square plots. Wight (1967) found that efficiency of rectangular plots dropped when the smaller dimension of the plot exceeded 1.5 m. Therefore, spatial distribution of shrubs and stems per shrub must be known to select appropriate plot size and shape.

Regression Equations. Regression equations for prediction of browse production from plant characteristics such as length and diameter of twigs would be more useful if these could be applied equally well throughout a large geographic area. However, this is not generally so because the magnitude of differences among years, seasons, and locations is usually great. Therefore, individual site–year equations should be used until such time that sufficient data are available for evaluation of site–regional equations.

A representative sample of the current year's twig growth must be obtained from the chosen site through a random sampling procedure to develop a regression equation for predicting browse production. Clip the twigs cleanly at the base and avoid browsed or broken twigs. Include twigs from all sides of the shrubs and from lower as well as upper branches. Obtain a range of twig lengths from minimum to maximum length. The clipped twigs should be placed in plastic bags to prevent desiccation until weighing.

6.3   FOREST BIOMASS

Tree, and subsequently, forest biomass is a much more important measure than volume of tree boles since all biomass materials are being used in much of the world. Yet, little information has been collected on other biomass components. Obtaining tree biomass estimates from correlated stem and canopy measurements follows the same procedures as those used for shrubs. Mensuration books specifically deal with volumetric estimates for timber saw-logs, and so forth. On the other hand, biomass estimates of stems, leaves, and bark often require more detailed measurements to adequately predict standing biomass of trees. Most procedures use statistical models that are solved by regression techniques.

6.3.1   Regression Models of Biomass Estimation

Plant ecologists should use statistical models to obtain estimates of tree biomass on an individual basis and sum over trees to obtain stand estimates. Components of biomass (stems, leaves, etc.) for trees can be estimated on an individual basis for stands. Stem diameter and tree height are easily measured, but secondary branches can present problems. Tritton and Hornbeck (1982) suggested that dbh, as a single measure, is sufficient to estimate biomass of the tree. Therefore, time is saved by this single measure that is easily obtained. Models used are simple and commonly include:

$$\text{Allometric} \quad y = ax^b \tag{6.33}$$

$$\text{Exponential} \quad y = ae^{bx} \tag{6.34}$$

$$\text{Quadratic} \quad y = a + bx + cx^2 \tag{6.35}$$

where y is the biomass component to be estimated; x is diameter and a, b, and c are regression coefficients. Baskerville (1972, 1965) and others discuss these models more fully.

Ouellet (1985) used a non-linear equation to predict individual total tree biomass for several species and for various components of individual trees:

stem wood, wood and bark of stem, and crown biomass (leaves and secondary branches). The model:

$$Y = B_1 D^{B_2} H^{B_3} + e \qquad (6.36)$$

can be used to predict the biomass of a given component Y, where the B_i ($i = 1, 2, 3$) are constants estimated by non-linear regression methods, and e is the error term. The model gave reasonably good results in predicting most component biomass for several species, as indicated by the percentage of variation accounted for (r^2) (Table 6.15).

A number of authors have reported results of predicting tree biomass by regressing biomass against various tree dimensions. Bole diameter, tree height, depth of crown, and crown diameters are commonly used in regression equations. To estimate the parameters of equations, individual trees must be dissected, measured, and biomass determined by component of interest. In general, the most successful equation is non-linear in form of the type given by Equation 6.33 or some variant in a log-transform model. If coefficients (parameters) are found to be numerically close in value for equations of several species, then a single equation can be used to predict biomass for several species. That is, by using measurements taken of the predictor values (s) from individual species, the equation to predict biomass can be solved.

Published equations can be used to derive preliminary estimates for components of the tree biomass. However, such equations are usually

TABLE 6.15 Coefficients of Determination R^2 for Predicting Oven-Dry Biomass (kg) of Total Above-Ground Tree and Components

Species	Total	Stem	Crown	MS[a]	MS wood	MS bark
Eastern white cedar (*Thuja occidentalis* L.)	.96	.98	.70	.97	.96	.94
Eastern hemlock (*Tsuga candensis* L. Carr.)	.98	.98	.88	.98	.97	.93
Red maple (*Acer rubra*)	.98	—	.82	.98	.98	.95
Beech (*Fagus grandifolia* Ehr.)	.98	—	.80	.98	.97	.91
Black ash (*Fraxinus nigra* Marsh.)	.97	—	.45	.97	.97	.95
White ash (*Fraxinus americana* L.)	.98	—	.79	.99	.99	.94

From Ouellet. 1985. Biomass equations for six commercial tree species in Quebec. *For. Chron.* 218–222.

[a]Total merchantable stem (MS) to 9 cm at top for conifers, to 9 cm in branches for deciduous species.

specific to both species and habitat characteristics and may not be useful if a high degree of precision is required. The use of regression equations to predict biomass of trees, like those of shrubs and herbaceous plants, are usually data specific. That is, equations are useful only within the range of data used to estimate model coefficients, and models should not be used to extrapolate predicted values outside of data used to formulate the model. In summary, the best overall estimate of tree biomass component or total may be a logarithmic function of dbh. For example, *log* (biomass) = *a* + *b log* (*dbh*).

6.3.2 Bark Biomass

Tree bark has a number of commercial uses, and bark production may be an important measurement to be made, along with other tree biomass components. It is often treated separately because equations usually are formulated to predict volume of bark rather than biomass, in which case a linear model may be used:

$$\log V_b = a + b \log HT + C \log dbh + d \log BT \tag{6.37}$$

where V_b is bark volume; HT is height; dbh is diameter at breast height; BT is double bark thickness at breast height; and a, b, c, and d are model coefficients (Kozak and Yang, 1981). Whittaker (1965, 1966) contributed much to the literature on equations derived for predicting tree component biomass from other tree measurements. Refer to Section 6.5 for more information on bark biomass estimation.

6.4 TREE FOLIAGE BIOMASS

Measurements of tree foliage biomass are needed to determine their contribution to the total productivity of a forest community. Foliage production can be used as a measure of evaluating site potential for forests because foliage biomass is determined by stand and site characteristics. In arid and semiarid regions of the world, trees are lopped for feeding foliage to animals, and materials are used as fuel for cooking. In addition, medium-sized trees, young trees, and coppice shoots are browsed by livestock as well as by large wildlife herbivores. Even large trees are within browsing range of camels and giraffes.

Production of foliage or browse of young trees and coppice shoots can be estimated by direct or harvest or nondestructive and indirect methods described earlier for measuring shrub production. However, in the case of tall trees, harvesting is difficult. Therefore, the most feasible method to estimate production of tree foliage is through indirect methods. That is, find some easily measured tree characteristics that are correlated with the biomass of leaves, stems, and branches. Crown diameter, foliage cover, basal area, diameter at breast height, diameter at the base of live crown and/or branches, tree height, height of crown alone or in combination, have been found to be good predictors of foliage biomass.

Reference Unit Technique. The reference or sample weight unit technique is used to estimate leaf biomass. Trees are first classified into groups based on compactness of the canopy foliage. Mason and Hutchings (1967) used three groups for Utah juniper (*Juniper utahensis*) (sparse, medium, and dense). A sample weight unit with the average foliage and fruit is selected and used as a standard for estimating fruit and foliage yields in each group. The number of weight units on each tree is then counted. Foliage and fruit are clipped from a sample unit, air-dried, and weighed. The total weight of the foliage and fruit on each tree is then estimated by multiplying the number of weight units by the weight of the sample unit. Estimates of mean foliage and fruit production for Utah juniper were predicted within 10% of the mean with 95% confidence for samples of 20 trees within each crown class. Each species would have to be studied for the appropriate unit used as a reference for the current year's growth. Sample size, needed to obtain sample adequacy, employs the standard sample size equation.

Crown Diameter. Tree foliage yield can be estimated from crown measurements. The correlation between crown and foliage production can be evaluated more effectively by inclusion of foliage denseness and soil characteristics measurements. Tree heights and crown diameters are measured to the nearest 0.15 m for each tree, and canopies of trees are classified into groups; for example, sparse, medium, and dense, based on the compactness of foliage. Mason and Hutchings (1967) estimated foliage production of Utah juniper from measurements of crowns on 400-m^2 randomly located plots. Logarithmic equations provided the best prediction equations. Correlation coefficients ranged from 0.88 to 0.98 for logarithmic equations relating foliage and fruit production to crown diameter. The equations, however, differed in coefficient values from site to site. Relations of current foliage yield of juniper-to-crown diameter for three foliage classes on five range sites are shown in Figure 6.9.

Equations relating weight of foliage and fruit to various tree measurements were developed by relating:

Logarithm of yield with logarithm of crown diameter
Logarithm of yield with logarithm of height
Logarithm of yield with logarithm of crown diameter and logarithm of height

Yield with height squared
Yield with height
Yield with crown diameter
Yield with crown diameter squared
Yield with crown surface
Yield with crown volume

Equations differed among sites so the site effect should be added to the model. Most investigators ignore such effects by developing an equation for each site. The use of a site term is easily incorporated and, in such cases, uses only one equation. For instance, the site effect is z in Eq. 6.8.

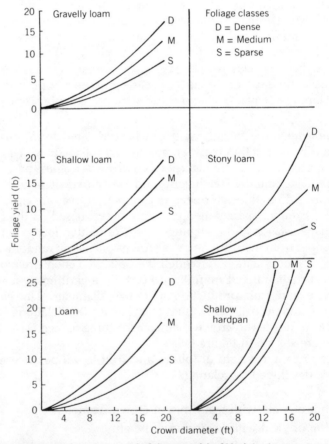

Figure 6.9 Relation of current year's foliage yield of Utah juniper to crown diameter for three foliage classes on five range sites (Redrawn from Mason and Hutchings. 1967 "Foliage yields." J. Range Manage. 20: 164.).

Crown Cover and Basal Area. Some studies have indicated that a high degree of correlation exists between leaf biomass and stem area at ground level or at breast height (1.4 m) and canopy area. Hutchings and Mason (1970) estimated foliage yields of gambel oak (*Quercus gambellii*) from foliage cover and basal area. Five to ten 20-m² square plots were used to sample dense stands of oak. Scattered stands were sampled by ten 40-m² plots. The number of stems per clump was recorded and foliage over 1.8 m in height was recorded separately. Sample weight units were clipped for foliage, acorns, and annual twig growth. Total air-dry weight, basal area, and foliage area were computed as follows:

Total air-dry weight = air-dry weight of samples × number of weight units in each clump

Basal area = average stem diameter² × number of stems

Foliage area = width of segment × length of segment

On most sites, foliage area was associated more closely with foliage production than either foliage volume or basal area. A multiple regression of foliage area and basal area provided the best prediction equation and accounted for over 75% of variation.

Stem and/or Branch Measurements. Periodic annual growth of stemwood, tree diameter, branch diameter, and circumference have also been found to have a high correlation with foliage growth and/or total biomass. The logic behind this correlation is that the amount of increment of stemwood or branchwood is a function of the amount of foliage which is photosynthesizing in a given period of time. Measurements of all first-order branches on sample trees, or on all trees on sample plots, can be used to estimate foliage biomass.

Several studies have found stem or branch diameter at the base of live crown as a better predictor of foliage biomass. All these measurements, except periodic annual growth, are easy to obtain. Kittredge (1944) found a well-defined linear trend when he plotted dry weight of jackpine (*Pinus* spp.) needles over the 5-year periodic annual increment in cubic feet. He also found a linear trend when leaf weight was plotted against stem diameter on double logarithmic paper. Rodionov (1959) obtained correlation coefficients of .90 to .93 between diameter at root collar and fresh weight of foliage for 2- and 4-year-old shelterbelts of oak (*Quercus* spp.) and ash (*Fraxinus* spp.) trees in Russia. He also found correlation coefficients of .89 and .86 for foliage weight with branch diameters for poplar (*Populus* spp.) and willow (*Salix* spp.), respectively. Elkington and Jones (1974) estimated fresh weights of leaves from regression of branch circumference on fresh weight of leaves per branch from

$$Y = 11.93 + 1.629X \qquad (6.38)$$

where Y represents leaf weight per branch and X represents branch circumferences. Measurements included several species of trees in Greenland, and they concluded that their model did not apply to other areas.

Elkington and Jones also suggested equations for predicting branch dry weight BW, leaf dry weight LW, and caudex dry weights CW based on a function of average branch circumference BC. In general form

$$BW^{1/3} = a_1 + b_1 (BC) \tag{6.39}$$

$$LW^{1/3} = a_2 + b_2 (BC) \tag{6.40}$$

$$CW^{1/3} = a_3 + b_3 \text{ (sum of } BC) \tag{6.41}$$

Note that the relationship is one of the cubic root of biomass and that the three biomass classes are linearly related to the branch circumference. Baskerville (1965) found, in balsam fir–white spruce–white birch (*Abies–Picea–Betula*) stands in New Brunswick, that total dry weight of foliage and branches of each species was closely correlated to stem dbh (r = .96 to .99).

Whisenant and Burzlaff (1978) and Felker et al. (1982) investigated the relationships between stem area and green and dry biomass of mesquite (*Prosopis* spp.) and other leguminous trees. Whisenant and Burzlaff found a significant linear relationship between stem area at 60 cm and green weight at different sites in Texas. The prediction equation over all the sites also had a high correlation coefficient (r = .93). The equation for predicting weight for these trees is of the simple form

$$Y = .410X \tag{6.42}$$

where Y is green weight of mesquite in kilograms per tree and X is the stem area at 60 cm above ground level. It was recommended that a minimum sample of 256 trees be measured when estimating tree weight to assure an acceptable precision level for the prediction model. This large number should indicate that stratification is needed for size classes, species, or other sources of variation. Felker et al. investigated stem diameter and height relationships to fresh and dry-weight biomass of tree legumes of *Prosposis*, *Cercidium*, *Olneya*, *Leucaena*, and *Parkinsonia* genera. They computed the following relationship to predict the biomass of the legume tree genera under study:

$$Y = .363X - .537 \tag{6.43}$$

where Y is the weight in kilograms per tree and X is the stem diameter. Obviously, for trees with diameter less than 1.5 cm, this relationship gives negative biomass. Yet log–log regression of basal stem diameter against biomass was solved for trees smaller than 1.5 cm in diameter.

6.5 CONSIDERATIONS FOR TREE BIOMASS SAMPLING

Annual Production of Tree Foliage. Annual production of foliage has been estimated from values of both standing biomass and litter fall. Estimates from standing biomass usually assume that annual production is equal to the weight of 1-year-old leaves on trees at the end of the growing season. This assumption overlooks any decline in weight of individual leaves at the end of the growing season. These losses are due to insect feeding and year-to-year fluctuations in foliage production. Hence, underestimates of total production are obtained. Foliage production estimates from litter traps involve similar assumptions and are also dependent on the type of litter trap employed. Whittaker (1966) based leaf production of deciduous species directly on foliage biomass and in evergreen trees, on a comparison of two estimates: foliage biomass (from regression on dbh) divided by years of leaf persistence, and from foliage production ratios to stem wood production. No correction was made for leaf consumption by animals, but an increase of the estimates of net production should be made to allow for unsampled plant parts such as bud-scales, flower parts, cones, and other fruits.

Production of Tree Biomass. Above-ground woody biomasses can be estimated by two independent approaches. One estimate is to sum stem wood and bark weight (Whittaker 1965) and normal foliage weights of deciduous and needleleaf evergreen forest (Whittaker 1962). The other estimate is obtained from total tree weights, and in some species branch and foliage weights, which are predicted from dbh values using regression equations already published for a species or genera.

Production estimates for tree strata can be based on the summation of the following [based on work by Whittaker (1966)]:

1. Stem wood growth is estimated by multiplying estimated volume increments by wood density values.
2. Stem production is estimated from the relation: $\Delta S/S = k(\Delta A/A)$, where S and ΔS are stem biomass and production, respectively; A and ΔA are basal area and basal area increment for a tree; and k is a constant.
3. Stem bark growth estimates are obtained by multiplying the stem wood growth estimate by the ratio of bark to stem wood basal area and multiplying this by an arbitrary correction for difference in relative growth rate.
4. Branch production is estimated from the equation $\Delta B/B = k(\Delta S/S)$, where ΔB and B are branch production and weight, and ΔS and S are wood production and weight. The separate estimates of branch, stem wood, and stem bark production should be checked against

      direct estimates of woody shoot production from the biomass and basal area increment ratio.

5. Leaf production of deciduous species is obtained directly from foliage biomass. Foliage production in evergreen species is based on two estimates: foliage biomass (from regression in dbh) divided by years of leaf persistence and foliage production from ratios to stem wood production.

6. Leaf, flower, and fruit loss is visually estimated. No correction was made for leaf consumption by animals, but the estimates of net production may be increased by a percentage to allow for unsampled plant parts such as bud-scales, corollas, cones, and other fruits.

7. Net shoot production estimates, then, are based on summing individual estimates for plant parts.

Biomass estimates are more reliable for stands of small trees for which appropriate regressions on stem diameter are available. However, such estimates can be affected by errors when equations are extrapolated to larger-sized trees. Estimates of biomass, made from parabolic volumes, are significantly affected by branch : stem ratios, while other estimates may be affected by actual : estimated volume increment ratios. These variables, then, should be studied carefully before they are used to predict tree biomass.

6.6 SELECTION OF SAMPLING UNITS FOR TREE BIOMASS

In the simple random selection process, every member of the population has equal probability of being selected. On the other hand, in systematic sampling, only the first unit is selected randomly and all subsequent units are equally spaced. If a vegetation type occurs in scattered clumps in the area to be sampled, then the number of sampling units in each clump should be proportional to the size of the clump, and in the literature this is referred to as "sampling with probability proportional to size" and abbreviated as "PPS sampling."

    In simple random or systematic sampling, all the sampling units are of the same size and shape. All trees in the sampling unit are measured for diameter and height classes and the class of trees is sampled in proportion to the number of trees in that class. In some forests, there are more small trees than large trees, and, thus, smaller trees would be sampled more intensively than bigger trees. Therefore, in uneven-aged forests, a sampling design that is based on probability of a tree being selected in a sampling unit proportional to the frequency of its size would give more precise estimates.

One type of probability sampling is 3P (probability proportional to prediction). This is a modification of the double-sampling procedure described earlier in this chapter. In the double-sampling procedure, the variables of interest are guessed on all sampling units n. On a subset of sample units n', the variables of interest are both guessed and measured. The subsample n' is randomly drawn from the larger sample size n. On the other hand, in 3P sampling, the variable of interest, usually tree volume, is guessed on all the trees in a stand. The estimated volume of the tree is compared with a randomly selected integer from a list. If the volume estimate is greater than or equal to the number, the tree is measured for accurate volume determination. The list has a set of integers from 1 through a maximum tree volume that can be expected. Further details on 3P or other similar sampling procedures are presented in textbooks on forest mensuration.

Sampling Procedures. Three main sampling procedures are used to determine biomass of tree components: unit area, the average tree, and regression analysis. The unit area method requires the collection of weights for tree components or foliage on a sample plot area or series of sample plots. The average tree method requires selection of trees considered to be of average biomass based on knowledge of linear dimensions of the trees. Foliage biomass per unit area is then determined by multiplication of the average number of trees in the unit area. This procedure, however, is not applicable in mixed-age forests. It is useful only where a rough estimate of total biomass is required. Baskerville (1965) reported that the average tree method could lead to gross errors in biomass estimation. The regression analysis method uses a mathematical relationship between weights of tree components and one or more of the easily measured tree dimensions. Regression equations may not be appropriate if sampling is restricted to a portion of population such as small trees or large trees. Results can be improved by stratifying the population into 5 or 10 size classes and selecting trees that are nearest to the average of each class.

Sampling error of random, stratified, and strip sampling designs was studied by Hasel (1938). Stratified sampling was more efficient than random sampling. The strips were tried because of clustering within strata. Systematic sampling within homogeneous strip plots was more efficient than random sampling. Random sampling and, to a greater extent, stratified random sampling of trees usually will yield residual mean squares about the regression line that are generally smaller than those found using all trees. Estimated standing weights for tree biomass from stratified sampling may be skewed by a few large overestimates. Stratification will reduce the number of trees required for a given precision. Variances and confidence intervals may be calculated using methods of Finney (1941) and Mountford and Bunce (1973). Madgwick and Satoo (1975) found that stratified random sampling gave smaller residual mean squares about regression than those found for random sampling or using all trees in plots.

Hasel studied relative efficiency of plots of varying sizes and shapes. The results of random sampling for size and shape of plot are given in Table 6.16. Among 4-ha plots, the long and narrow shape was most efficient. Variations in forests are usually greatest at right angles to contour lines, and sampling at right angles to the contours with long, narrow plots is more advantageous.

Bormann (1953) conducted a study on effective size and shape of plots in the Piedmont plateau of North Carolina. An area of 140 m on a side was divided into 4900 2 × 2-m plots. Basal area was determined for two distributions: (1) one containing 82.5% of the total basal area and represented the type vegetation and (2) one containing only 11% of the total basal area sporadically distributed over the area. Thus, each plot had a value for each distribution associated with it. Plots were combined in various sizes, shapes, and orientations, and the variances were calculated. Long, narrow plots for the first distribution were most efficient when the main axis ran across contours or soil banding. Variances decreased with plot length, but the smallest variances were obtained with wider plots. There was little difference in variance for different plot orientations. Variances increased with length if plots paralleled the contours in the second distribution but decreased if the orientation was across contours. That is, an increased edge affects errors for long narrow plots. Longer plots were considered easier to randomly locate in the field than were shorter plots.

The following four designs would be best for the first kind of distribution: 4 × 140 m (B), 10 × 70 m (A), 10 × 140 m (B), and 10 × 140 m (A). The second distribution was best sampled by 4 × 140 m (B) and 10 × 140 m (B) plots. Type A assumes that the long axis is parallel to the direction of greatest variation, and type B is perpendicular to that variation.

Plot Size and Shape. The basic considerations in selecting an optimum plot size and shape are the same as discussed in Section 6.1.5 for measurement

TABLE 6.16 Relative Efficiency of Plots of Varying Size and Shape

Plot Size and Shape	Mean Square	Efficiency (%)	Relative Size of Sample for a Given Precision of Estimate
1 ha: 50 × 200 m	282.4961	100.00	1.00
2 ha: 100 × 200 m	449.3926	62.86	1.59
4 ha: 50 × 800 m	566.0602	49.90	2.00
100 × 400 m	711.8291	39.68	2.52
200 × 200 m	687.4958	41.09	2.43

From Hasel. 1938. Sampling errors in timber surveys. *J. Agric. Res.* **57**: 719.

of herbaceous vegetation. In forest inventory, the sampling units are either in the shapes of strips or circular plots. The strips are usually 20 m wide and run from one edge to the other edge of the boundary. The wide strip is reduced to 10 m in thick growth and increased to 40 m if the trees are widely scattered. The circular plots are usually 11–12 ha in North America and 0.01–0.05 ha in Europe. Although circular plots are more common, some authors have used square and rectangular plots.

The only advantage of strip sampling is that time is saved in traveling from plot to plot. However, an optimum size of plot is one in which equal amounts of time are spent on travel and on plot measurement (Zeide 1980). Zeide (1980), O'Regan and Arvanitis (1966), and Tardif (1965) provide further details on optimum size and shape of plots for tree measurements.

6.7 BIBLIOGRAPHY

Ahmed, J., and C. D. Bonham. 1982. Optimum allocation in multivariate double sampling for biomass estimation. *J. Range Manage.* **35:** 777–779.

Ahmed, J., C. D. Bonham, and W. A. Laycock. 1983. Comparison of techniques used for adjusting biomass estimates by double sampling. *J. Range Manage.* **36:** 217–221.

Alemdag, I. S., and G. M. Bonnor. 1985. Biomass inventory of federal forest lands at Petawawa: A case study. *Forestry Chron.* **61:** 81–86.

Alemdag, I. S., and K. W. Horton. 1981. Single-tree equations for estimating biomass of trembling aspen, large tooth aspen, and white birch in Ontario. *Forestry Chron.* **57:** 169–173.

American Society of Agronomy, American Dairy Science Association, American Society of Animal Production, and American Society of Range Management. 1952. Pasture and range research techniques. *Agron. J.* **44:** 39–50.

Andariese, S. W., and W. W. Covington. 1986. Biomass estimation for four common grass species in northern Arizona ponderosa pine. *J. Range Manage.* **39:** 472–473.

Anderson, D. M., and M. M. Kothmann. 1982. A two-step sampling technique for estimating standing crop of herbaceous vegetation. *J. Range Manage.* **35:** 675–677.

Andrew, M. H., I. R. Noble, and R. T. Lange. 1979. A non-destructive method for estimating the weight of forage on shrubs. *Aust. Range J.* **1:** 225–231.

Andrew, M. H., I. R. Noble, R. T. Lange, and A. W. Johnson. 1981. The measurement of shrub forage weight: Three methods compared. *Aust. Range J.* **3:** 74–82.

Barrett, J. P., and W. A. Guthrie. 1969. Optimum plot sampling in estimating browse. *J. Wildl. Manage.* **33:** 399–403.

Bartolome, J. W., and B. H. Kosco. 1982. Estimating browse production by deer-shrub (*Ceonothus integerrimus*). *J. Range Manage.* **35:** 671–672.

Basile, J. V., and S. S. Hutchings. 1966. Twig diameter–length–weight relations of bitterbrush. *J. Range Manage.* **19:** 34–38.

Baskerville, G. L. 1965. Estimation of dry weight of tree components and total standing crop in conifer stands. *Ecology* **46:** 867–869.

——. 1972. Use of logarithmic regression in the estimation of plant biomass. *Can. J. Forestry* **2:** 49–53.

Beauchamp, J. J., and J. S. Olson. 1973. Correction for bias in regression estimates after logarithmic transformation. *Ecology* **54:** 1403–1407.

Bentley, J. R., D. W. Seegrist, and D. A. Blakeman. 1970. *A Technique for Sampling Low Shrub Vegetation by Crown Volume Classes.* USDA Forest Service Research Note PSW-215, 11 pp.

Blackman, T. 1978. Bark pellets are high-energy fuel for coal, gas application. *Forestry Ind.* **105:** 18–19.

Blair, R. M. 1959. Weight techniques for sampling browse production on deer ranges, pp. 26–31. In *Techniques and Methods of Measuring Understory Vegetation.* Proceedings of Symposium at Tifton, GA, October 1958, USDA, 174 pp.

Blankenship, J. O., and D. R. Smith. 1966. Indirect estimation of standing crop. *J. Range Manage.* **19:** 74–77.

Bobek, B., and R. Dzieciolowski. 1972. Method of browse estimation in different types of forests. *Acta Theriologica* **17:** 171–186.

Bonham, C. D. 1984. *A Methodology to Determine the Vertical Distribution of Forage Biomass on Rangelands of Kenya: Principles and Practices.* Winrock International, Morrilton, AR. Technical Report No. 2.

Bonham, C. D., L. L. Larson, and A. M. Morrison. 1980. *A Survey of Techniques for Measurement of Herbaceous and Shrub Production, Cover and Diversity on Coal Lands in the West.* Office of Surface Mining Region V, January 1980.

Bormann, F. H. 1953. The statistical efficiency of sample plot size and shape in forest ecology. *Ecology* **34:** 474–487.

Boutton, T. W., and L. L. Tieszen. 1983. Estimation of plant biomass by spectral reflectance in an East African grassland. *J. Range Manage.* **36:** 213–216.

Brown, D. 1954. *Methods of Surveying and Measuring Vegetation.* Bulletin No. 42, Commonwealth Bureau of Pastures and Field Crops, Hurley, Berkshire, 223 pp.

Brown, J. K. 1976. Estimating shrub biomass from basal stem diameters. *Can. J. Forestry Res.* **6:** 153–158.

Bryant, F. C., and M. M. Kothman. 1979. Variability in predicting edible browse from crown volume. *J. Range Manage.* **32:** 144–146.

Campbell, R. S., and J. T. Cassady. 1949. Determining forage weight on southern forest ranges. *J. Range Manage.* **2:** 30–32.

Campbell, R. S., and J. T. Cassady. 1955. *Forage Weight Inventories on Southern Forest Ranges.* U.S. Forest Service, Southern Forest Experiment Station Occasional Paper 139, 18 pp.

Carhart, A. H., and H. Means. 1941. Forage weight per acre method for appraisal of herbiverous animal requirements. *Colo. Game Fish Comm., Pittman–Robertson Deer–Elk Surv.* **5:** 29–51.

Christidis, B. G. 1931. The importance of the shape of plots in field experimentation. *J. Agric. Sci.* **21:** 14–37.

Clark, I. 1945. Variability in growth characteristics of forage plants on summer range in central Utah. *J. Forestry* **43**: 273–283.

Cochran, W. G. 1977. *Sampling Techiques*, 3rd ed. Wiley, New York, 428 pp.

Crafts, E. C. 1938. Height–volume distribution in range grasses. *J. Forestry* **36**: 1182–1185.

———. 1939. Measuring the annual forage crop. U.S. Forest Service, *Proceedings of Range Research Seminar*, pp. 143–147.

Currie, P. O., and G. Peterson. 1966. Using growing season precipitation to predict crested wheatgrass yields. *J. Range Manage.* **19**: 284–288.

Currie, P. O., M. J. Morris, and D. L. Neal. 1973. Uses and capabilities of electronic capacitance instruments for estimating standing herbage. *J. Br. Grassl. Soc.* **28**: 155–160.

Czarnowski, M. S. 1973. On the map and model of productive site-capacity on the Earth (in Polish, with English summary). *Geogr. Rev.* **45**: 295–308.

Dalkes, P. D. 1941. Use and availability of the more common browse plants in the Missouri Ozarks. *Transactions of Sixth North American Wildlife Conference*, pp. 155–160.

Daubenmire, R. F. 1968. *Plant Communities: A Textbook of Plant Synecology.* Harper & Row, New York, 300 pp.

Davis, J. G. 1931. *The Experimental Error of the Yield from Small Plots of Natural Pasture.* Australian Council on Science and Industry Research Bulletin No. 48.

Davies, W. 1931. *Methods of Pasture Analysis and Fodder Sampling.* Welsh Plant Breeding Station, Aberystwyth, Report No. 1, 29 pp.

Davis, J. B., P. T. Tueller, and A. D. Bruner. 1972. Estimating forage production from shrub ring widths in Hot Creek Valley, Nevada. *J. Range Manage.* **25**: 398–402.

Dimitrov, E. T. 1976. Mathematical models for determining the bark volume of spruce in relation to certain mensurational characteristics. *Forestry Abstr.* **37**: 6281.

Duncan, D. A., and R. G. Woodmansee. 1975. Forecasting forage yield from precipitation in California's annual rangelands. *J. Range Manage.* **28**: 327–329.

Earle, D. F., and A. A. McGowan. 1979. Evaluation and calibration of an automated rising plate meter for estimating dry matter yield of pasture. *Aust. J. Exp. Agric. Anim. Husb.* **19**: 337–343.

Elkington, T. T., and B. M. G. Jones. 1974. Biomass and primary productivity of birch (*Betula pubescens* S. LAT.) in S.W. Greenland. *J. Ecol.* **62**: 821–830.

Evans, F. C. 1952. *The Influence of Size of Quadrat on the Distributional Patterns of Plant Populations.* (Contributions from the Laboratory of Vertebrate Biology of the University of Michigan.) 54, 15 pp.

Felker, P., P. R. Clark, J. F. Osborn, and G. H. Cannell. 1982. Biomass estimation in a young stand of Mesquite (*Prosopis* spp.), Ironwood (*Olneya tesota*), Palo Verde (*Cercidium floridium*), and (*Parkinsonia aculeata*), and Leucaena (*Leucaena leucocephala*). *J. Range Manage.* **35**: 87–89.

Ferguson, R. B., and M. A. Marsden. 1977. Estimating overwinter bitterbrush utilization from twig diameter-length-weight relations. *J. Range Manage.* **30**: 231–236.

Finney, D. J. 1941. On the distribution of a variate whose logarithm is normally distributed. *J. Roy. Stat. Soc., Ser. B* **2:** 155–161.

Ford, E. D., and P. J. Newbould. 1977. The biomass and production of ground vegetation and its relation to tree cover through a deciduous woodland cycle. *J. Ecol.* **65:** 201–212.

Gillen, R. L., and E. L. Smith. 1986. Evaluation of the dry-weight-rank method for determining species composition in tallgrass prairie. *J. Range Manage.* **39:** 283–285.

Goebel, C. J., L. DeBano, and R. D. Lloyd. 1958. A new method of determining forage cover and production on desert shrub vegetation. *J. Range Manage.* **11:** 244–246.

Grosenbaugh, L. R. 1967. The gains from sample-tree selection with unequal probabilities. *J. Forestry* **65:** 203–206.

Gysel, L. W. 1957. Effects of silvicultural practices on wildlife food and cover in oak and aspen types in northern Michigan. *J. Forestry* **55:** 803–809.

Halls, L. K., and F. R. Harlow. 1971. Weight–length relations in flowering dogwood twigs. *J. Range Manage.* **24:** 236–237.

Harniss, R. O., and R. B. Murray. 1976. Reducing bias in dry leaf weight estimates of big sagebrush. *J. Range Manage.* **29:** 430–432.

Hasel, A. A., 1938. Sampling error in timber surveys. *J. Agric. Res.* **57:** 713–736.

Heady, H. F. 1950. Studies on bluebunch wheatgrass in Montana and height–weight relationships of certain range grasses. *Ecol. Monogr.* **20:** 55–81.

Heady, H. F., and D. T. Torell. 1959. Forage preference exhibited by sheep with esophageal fistulas. *J. Range Manage.* **12:** 28–34.

Heady, H. F., and G. M. Van Dyne. 1965. Prediction of weight composition from point samples on clipped herbage. *J. Range Manage.* **18:** 144–148.

Hickey, W. C., Jr. 1961. Relation of selected measurements to weight of crested wheatgrass plants. *J. Range Manage.* **14:** 143–146.

Hilmon, J. B. 1959. Determination of herbage weight by double-sampling: weight estimate and actual weight, pp. 20–25. In *Techniques and Methods of Measuring Understory Vegetation.* Proceedings of Symposium at Tifton, GA, October 1958, USDA, 174 pp.

Holmes, C. W., 1974. The Massey grass meter. *Dairy Farming Annual* (1974) 26–30.

Hughes, E. E. 1962. Estimating herbage production using inclined point frame. *J. Range Manage.* **15:** 323–325.

Hughes, R. H. 1959. The weight-estimate method in herbage production determinations, pp. 17–19. In *Techniques and Methods of Measuring Understory Vegetation.* Proceedings of Symposium at Tifton, GA, October 1958, USDA, 174 pp.

Hutchings, S. S., and L. R. Mason. 1970. Estimating yields of gambel oak from foliage cover and basal area. *J. Range Manage.* **23:** 430–434.

Jones, R. M., and J. N. G. Hargreaves. 1979. Improvements to the dry-weight-rank method for measuring botanical composition. *Grass Forage Sci.* **34:** 181–189.

Kelly, A. F. 1958. A comparison between two methods of measuring seasonal growth of two strains of *Dactylis glomerata* when grown as spaced plants and in swards. *Br. Grassl. Soc. J.* **13:** 99–105.

Kelly, J. M., G. M. Van Dyne, and W. F. Harris. 1974. Comparison of three methods of assessing grassland productivity and biomass dynamics. *Am. Midl. Nat.* **92:** 357–369.

Kennedy, R. K. 1972. Preliminary network evaluation on methods of primary producer biomass estimation, pp. 30–46. In *Preliminary Producer Data Synthesis, 1970. Comprehensive Sites,* P. G. Risser (ed.). US/IBP Grassland Biome Technical Report No. 161, Colorado State University, Fort Collins.

Kinsinger, F. E., and G. S. Strickler. 1961. Correlation of production with growth and ground cover of whitesage. *J. Range Manage.* **14:** 274–278.

Kirmse, R. D., and B. E. Norton. 1985. Comparison of the reference unit method and dimension analysis methods for two large shrubby species in the Caatinga Woodlands. *J. Range Manage.* **38:** 425–428.

Kittredge, J. 1944. Estimation of the amount of foliage of trees and stands. *J. Forestry* **42:** 905–912.

———. 1945. Some quantitative relations of foliage in the chaparral. *Ecology* **26:** 70–73.

Koch, E. J., and J. A. Rigney. 1951. A method of estimating optimum plot size from experimental data. *Agron. J.* **43:** 17–21.

Kozak, A., and R. C. Yang. 1981. Equations for estimating bark volume and thickness of commercial trees in British Columbia. *Forestry Chron.* **57:** 112–115.

Lauenroth, W. K. 1970. Dynamics of dry-matter production in a mixed grass prairie in western North Dakota. M.S. Thesis, North Dakota State University, Fargo, 102 pp.

Lay, D. W. 1956. Effects of prescribed burning on forage and mast production in southern pine forests. *J. Forestry* **54:** 582–584.

Le Houerou, H. N., and C. H. Hoste. 1977. Rangeland production and annual rainfall relations in the Mediterranean Basin and in the African Sahelo–Sudanian Zone. *J. Range Manage.* **30:** 181–189.

Leith, H. 1973. Primary production: Terrestrial ecosystems (paper presented at 2nd IBS Congress, October 1971, Miami, FL). *Human Ecol.* **1:** 303–332.

Leith, H., and E. Box. 1972. Evapotranspiration and primary productivity: C. W. Thornthwaite memorial model, pp. 37–46. In *Papers on Selected Topics in Climatology,* J. R. Mather (ed.). Thornthwaite Associates, Elmer, NJ.

Lester, J. 1969. Net shoot production and biomass transfer rates in a mature grassland ecosystem. M.S. Thesis, Fort Hays Kansas State College, Hays, 26 pp.

Lomnicki, A., E. Brandola, and K. Jankowska. 1968. Modification of the Wiegert–Evans method for estimation of net primary production. *Ecology* **49:** 147–149.

Ludwig, J. A., J. F. Reynolds, and P. D. Whitson, 1975. Size–biomass relationship of several chihuahuan desert shrubs. *Am. Midl. Nat.* **94:** 451–461.

Lundblad, K. 1937. Methods for botanical analysis of grassland. Svenska Mosskulturfor. *Tidskr.* (Sweden) **51:** 187–235.

Lyon, L. J. 1968. Estimating twig production of serviceberry from crown volumes. *J. Wildl. Manage.* **32:** 115–119.

Madgwick, H. A. I., and T. Satoo. 1975. On estimating the aboveground weights of tree stands. *Ecology* **56:** 1446–1450.

Mall, L. P., and S. K. Billore. 1974. Production relations with density and basal area in grassland community. *Geobios* **1**: 84–86.

Mall, L. P., and R. Tugnawat. 1973. Biomass estimation of producers in grassland ecosystems. *Current Sci.* **42**: 868–869.

Malone, C. R. 1968. Determination of peak standing crop biomass of herbaceous shoots by the harvest method. *Am. Midl. Nat.* **79**: 429–435.

't Mannetje, L., and K. P. Haydock. 1963. The dry-weight-rank method for the botanical analysis of pasture. *J. Br. Grassl. Soc.* **18**: 268–275.

Mason, L. R., and S. S. Hutchings. 1967. Estimating foliage yields on Utah juniper from measurements of crown diameter. *J. Range Manage.* **20**: 161–166.

Medin, D. E. 1960. Physical site factors influencing annual production of true mountain mahogany (*Cercocarpus montanus*). *Ecology* **41**: 454–460.

Meyer, H. A. 1946. Bark volume determination in trees. *J. Forestry* **44**: 1067–1070.

Michell, P. and R. V. Large. 1983. The estimation of herbage mass of perennial ryegrass sward: A comparative evaluation of a rising-plate meter and a single

Mitchell, P. and R. V. Large. 1983. The estimation of herbage mass of perennial ryegrass sward: A comparative evaluation of a rising-plate meter and a single probe capacitance meter calibrated at and above ground level. *Grass Forage Sci.* **38**: 295–299.

Morris, M. J. 1959. *Some Statistical Problems in Measuring Herbage Production and Utilization*. Symposium on Techniques and Methods of Measuring Understory Vegetation. Tifton, GA, 1958. Publication of USDA Forest Service, pp. 139–145.

Moszynska, B. 1970. Estimation of the green top production of the herb layer in a bog pinewood *Vaccinis uliginosi-pinetum*. *Ekologia Polska* **18**: 779–803.

Mountford, M. D., and R. G. H. Bunce. 1973. Regression sampling with allometrically related variables with particular reference to production studies. *Forestry* **46**: 203–212.

Mueller-Dombois, D., and H. Ellenberg. 1974. *Aims and Methods of Vegetation Ecology*. Wiley, New York, 547 pp.

Murray, R. B., and M. Q. Jacobson. 1982. An evaluation of dimension analysis for predicting shrub biomass. *J. Range Manage.* **35**: 451–454.

Odum, E. P. 1960. Organic production and turnover in old field succession. *Ecology* **41**: 34–39.

Ouellet, D. 1985. Biomass equations for six commercial tree species in Quebec. *Forestry Chron.* **61**: 218–222.

O'Regan, W. G., and L. G. Arvanitis. 1966. Cost effectiveness in forest sampling. *Forest Sci.* **12**: 406–414.

Ovington, J. D., D. Heitkamp, and D. B. Lawrence. 1963. Plant biomass and productivity of prairie, savannah, oakwood and maize field ecosystems in central Minnesota. *Ecology* **44**: 52–63.

Papanastasis, V. P. 1977. Optimum size and shape of quadrat for sampling herbage weight in grasslands of northern Greece. *J. Range Manage.* **30**: 446–449.

Pasto, J. K., J. R. Allison, and J. B. Washko. 1957. Ground cover and height of sward as a means of estimating pasture production. *Agron. J.* **49**: 407–409.

Payandeh, B. 1981. Some applications of nonlinear regression models in forestry research. *Forestry Chronicle* **59**: 244–248.

Payne, G. F. 1974. Cover–weight relationships. *J. Range Manage.* **27**: 403–404.

Pearse, K. 1935. An area-list method of measuring range plant populations. *Ecology* **16**: 573–579.

Pearson, R. L., and L. D. Miller. 1972a. *Remote Spectral Measurements as a Method for Determining Plant Cover.* US/IBP Grassland Biome Technical Report No. 167, Colorado State University, Fort Collins, 45 pp.

————, R. L., and L. D. Miller. 1972b. Remote mapping of standing crop biomass for estimation of the productivity of the shortgrass prairie, Pawnee National Grasslands, Colorado, pp. 1357–1381. In *Proceedings of the 8th International Symposium on Remote Sensing Environment,* University of Michigan, Ann Arbor.

Pechanec, J. F., and G. D. Pickford. 1937. A weight estimate method for determination of range or pasture production. *J. Am. Soc. Agron.* **29**: 894–904.

Pechanec, J. F., and G. Stewart. 1940. Sagebrush–grass range sampling studies: Size and structure of sampling units. *J. Am. Soc. Agron.* **32**: 669–682.

Precsenyi, S. 1957. The correlation between ground-cover and vegetation yield. *Portug. Acta Biol., Ser. B.* **6**: 94–96.

Provenza, F. D., and P. J. Urness. 1981. Diameter-length–weight relations for blackbrush (*Coleogyne ramosissima*) branches. *J. Range Manage.* **34**: 215–217.

Reppert, J. N., R. H. Hughes, and D. A. Duncan. 1962. Herbage yield and its correlation with other plant measurements, pp. 15–21. In *Range Research Methods.* USDA Miscellaneous Publication No. 940, 172 pp.

Rittenhouse, L. R., and F. A. Sneva. 1977. A technique for estimating big sagebrush production. *J. Range Manage.* **30**: 68–70.

Robel, R., J. Briggs, A. Dayton, and L. Hulbert. 1970. Relationships between visual obstruction measurements and weight of grassland vegetation. *J. Range Manage.* **23**: 295–297.

Rodionov, M. S. 1959. Determining foliage weight in shelter belts. *Bot. Zuhr. SSSR* (USSR) **44**: 333–337.

Rosenzweig, M. L. 1968. Net primary productivity of terrestrial communities: Prediction from climatological data. *Am. Nat.* **102**: 67–74.

Roussopoulos, P. J., and R. M. Loomis. 1979. *Weights and Dimensional Properties of Shrubs and Small Trees of the Great Lakes Conifer Forest.* USDA Forest Service Research Paper. NC-178, North Central Forest Experiment Station, 6 pp.

Ruark, G. A., G. L. Martin, and J. G. Bockheim. 1987. Comparison of constant and variable allometric ratios for estimating *Populus tremuloides* biomass. *Forest Sci.* **33**: 294–300.

Sandland, R. L., J. C. Alexander, and K. P. Haydock. 1982. A statistical assessment of the dry-weight rank method of pasture sampling. *Grass Forage Sci.* **37**: 263–272.

Schneider, M. H. 1977. Energy from forest biomass. *Forestry Chron.* **63**: 215–218.

Schuster, J. L. 1965. Estimating browse from twig and stem measurements. *J. Range Manage.* **18**: 220–222.

Scott, D. 1986. Coefficients for the dry-weight rank method of botanical analysis of pasture. *Grass Forage Sci.* **41:** 319–321.

Shafer, E. L. 1963. The twig-count method for measuring hardwood deer browse. *J. Wildl. Manage.* **27:** 428–437.

Singh, J. S. 1968. Net aboveground community productivity in the grasslands at Varsani, pp. 631–654. In *Proceedings of the Symposium on Recent Advances in Tropical Ecology*, R. Misra and B. Gopal (eds.). ISTE, Varsani, India.

Singh, J. S., and P. S. Yadava. 1972. Biomass structure and net primary productivity in the grassland ecosystem at Kurukshetra, pp. 59–74. In *Papers from a Symposium on Tropical Ecology with an Emphasis on Organic Productivity*, P. M. Golley and F. B. Golley (eds.). University of Georgia, Athens.

Singh, J. S., and P. S. Yadava. 1974. Seasonal variation in composition, plant biomass, and net primary productivity of a tropical grassland at Kurukshetra, India. *Ecol. Monogr.* **44:** 351–376.

Singh, J. S., W. K. Lauenroth, and R. K. Steinhorst. 1975. Review and assessment of various techniques for estimating net aerial primary production in grasslands from harvest data. *Bot. Rev.* **41:** 181–232.

Smith, H. F. 1938. An empirical law describing heterogeneity in the yields of agricultural crops. *J. Agric. Sci.* **28:** 1–23.

Sneva, F., and D. N. Hyder. 1962. Estimating herbage production on semiarid ranges in the Intermountain Region. *J. Range Manage.* **15:** 88–93.

Stanek, W., and D. State. 1978. *Equations Predicting Primary Productivity (Biomass) of Trees, Shrubs, and Lesser Vegetation based on Current Literature.* Canada Pacific Forestry Research Center, Information Report BC-X-183. 58 pp.

Stayton, C. L., and M. Hoffman. 1970. *Estimating Sugar Maple Bark Thickness and Volume.* USDA Forest Service Research Paper, North Central Forest Experiment Station NC-38, 7 pp.

Stockdale, C. R., and K. B. Kelly. 1984. A comparison of a rising-plate meter and an electronic capacitance meter for estimating the yield of pastures grazed by dairy cows. *Grass Forage Sci.* **39:** 391–394.

Sukhatme, P. V. 1947. The problem of plot size in large-scale yield surveys. *J. Am. Stat. Assoc.* **42:** 297–310.

Tardif, G. 1965. Some considerations concerning the establishment of optimum plot size in forest survey. *Forestry Chron.* **41:** 93–102.

Teare, I. D., and G. O. Mott. 1965. Estimating forage yield in situ. *Crop Sci.* **5:** 311–313.

Thompson, H. R. 1958. The statistical study of plant distribution patterns using a grid of quadrats. *Aust. J. Bot.* **6:** 322–342.

Toledo, J. M., J. C. Burns, H. L. Lucas, Jr., and A. Angelone. 1980. Herbage measurement *in situ* by electronics. 3. Calibration, characterization, and field application of the earth-plate forage capacitance meter: A prototype. *Grass Forage Sci.* **35:** 189–196.

Torrie, J. H., D. R. Schmidt, and G. H. Tenpas. 1963. Estimates of optimum plot size and shape and replicate number for forage yield of alfalfa–bromegrass mixtures. *Agron. J.* **55:** 258–260.

Tritton, L. M., and J. W. Hornbeck. 1982. *Biomass Equations for Major Tree Species of Northeast*. USDA Forest Service General Technical Paper NE-69, Northeast Forest Experiment Station, 46 pp.

Tucker, C. J. 1980. A critical review of remote sensing and other methods for nondestructive estimation of standing crop biomass. *Grass Forage Sci.* **35**: 177–182.

Uresk, D. W., R. O. Gilbert, and W. H. Rickard. 1977. Sampling big sagebrush for phytomass. *J. Range Manage.* **30**: 311–314.

Van Dyne, G. M., W. G. Vogel, and H. G. Fisser. 1963. Influence of small plot size and shape on range herbage production estimates. *Ecology* **44**: 746–759.

Vickery, P. J., I. L. Bennett, and G. R. Nicol. 1980. An improved electronic capacitance meter for estimating herbage mass. *Grass Forage Sci.* **35**: 247–252.

Wallentinus, H. G. 1973. Above-ground primary production of a *Juncetum gerardi* on a Baltic sea-shore meadow. *Oikos* **24**: 200–219.

Waller, S. S., M. A. Brown, and J. K. Lewis. 1981. Factors involved in estimating green biomass by canopy spectroreflectance measurements. *J. Range Manage.* **34**: 105–108.

Wassom, C. E., and R. R. Kalton. 1953. *Estimations of Optimum Plot Size Using Data from Bromegrass Uniformity Trials*. Iowa State College Research Bulletin 396.

Wells, K. F. 1971. Measuring vegetation changes of fixed quadrats by vertical ground stereophotography. *J. Range Manage.* **26**: 233–236.

Westell, C. E., Jr. 1954. Available browse following aspen logging in lower Michigan. *J. Wildl. Manage.* **18**: 266–271.

Whisenant, S. G., and D. F. Burzlaff. 1978. Predicting green weight of mesquite (*Prosopis glandulosa* Torr.). *J. Range Manage.* **31**: 396–397.

Whittaker, R. H. 1962. Net production relations of shrubs in the Great Smoky Mountains. *Ecology* **43**: 357–377.

———. 1965. Branch dimensions and estimation of branch production. *Ecology* **46**: 365–370.

———. 1966. Forest dimensions and production in the Great Smoky Mountains. *Ecology* **47**: 103–121.

Wiegert, R. G. 1962. The selection of an optimum quadrat size for sampling the standing crop of grasses and forbs. *Ecology* **43**: 125–129.

Wiegert, R. G., and F. C. Evans. 1964. Primary production and disappearance of dead vegetation on an old field in southeastern Michigan. *Ecology* **45**: 49–63.

Wight, J. R. 1967. The sampling unit and its effect on saltbush yield estimates. *J. Range Manage.* **20**: 323–325.

Wight, J. R., and E. C. Neff. 1983. *Soil–Vegetation–Hydrology Studies*, Vol. II, *User Manual for ERHYM: The Ekalaka Rangeland Hydrology and Yield Model*. USDA, ARS, Agriculture Research Service, ARR-W-29.

Wight, J. R., C. L. Hansen, and D. Whitmer. 1984. Using weather records with a forage production model to forecast range forage production. *J. Range Manage.* **37**: 3–6.

Wilm, H. G., D. F. Costello, and G. E. Kipple. 1944. Estimating forage yield by the double-sampling method. *J. Am. Soc. Agron.* **36**: 194–203.

Wimbush, D. J., M. D. Barrow, and A. B. Costin. 1967. Color stereophotography for measurement of vegetation. *Ecology* **48**: 150–152.

Wisiol, K. 1984. Estimating grazingland yield from commonly available data. *J. Range Manage.* **37**: 471–475.

Woodwell, G. M., and R. H. Whittaker. 1968. Primary production in terrestrial ecosystems. *Am. Zoologist* **8**: 19–30.

Zeide, B. 1980. Plot size optimization. *Forest Sci.* **26**: 251–257.

7 MONITORING AND EVALUATION

Monitoring and evaluation of vegetation is the key to natural resource management. The need for proper resource planning in the United States resulted in the enactment of the first major legislation of the Forest and Rangelands Renewable Resources Planning Act of 1974 (RPA). The RPA emphasized inventory and assessment of current resources and long-range planning. To accomplish vegetation resource planning, the first step is to quantify vegetation characteristics, and the second step is to make an estimate of any change that may occur in vegetation over time. The last step of vegetation assessment is to project trends in vegetation characteristics into the future (Buckman and Van Sickle, 1983).

Monitoring and evaluation of vegetation characteristics require a database of both soil and vegetation resources. Classification of vegetation into homogeneous units should be made according to vegetation types, ecological sites, or range sites. Vegetation types are usually formed on the basis of one or two species that are dominant or codominant. Dominance is usually determined on the basis of species size and abundance. Site characteristics are then often described after vegetation has been characterized. An ecological site is a unit of land with specific physical characteristics and has the potential to support a specific plant association. Several ecological sites may be found within a given vegetation type. Terms such as "range site" and "habitat type" are also frequently used in the literature. The two latter terms are similar but are not identical in concept. The term "ecological site" embodies the concept of both "range site" and "habitat type" and may be preferred by fieldworkers. All discussion in this chapter applies to vegetation types, range sites, and ecological sites.

7.1 MAPPING UNITS

There are several ways to prepare a vegetation map showing ecological sites. It is beyond the scope of this book to describe in detail the cartographic techniques for vegetation mapping; however, a brief overview of the technique is described. One should be able to accomplish the compilation of an accurate vegetation map by using topographic maps, soil survey maps, and aerial photographs. Color photographs provide detailed information on both soil and vegetation and are quite useful for detailed vegetation mapping.

The first step is to delineate general vegetation types on aerial photographs. Each vegetation type has a characteristic signature based on the tone and texture portrayed by photographs. A reconnaissance field survey is then undertaken to become familiar with different vegetation types. During the survey, a list is made of important plant species occurring within the area, and descriptions are made of topography, aspect and slope. Boundaries of vegetation types, as they occur on the ground, are delineated on the topographic survey maps. Soil survey maps are useful in delineating the boundaries of vegetation types because such boundaries often follow those of soil types.

The second step is to compare overlays of the vegetation boundaries delineated on aerial photographs to those on topographic sheets. Discrepancies, if any, are corrected through an additional field check. The field-produced boundaries are generally more accurate with respect to correct identification of vegetation types. On the other hand, the photo-produced overlays are generally more accurate with respect to the overall pattern and distribution of each vegetation type. After the field check, a composite overlay is prepared by putting together the pieces of corrected overlays of the topographic maps.

7.1.1 Naming Vegetation Units

Vegetation units should be given a name that represents the abiotic and/or biotic features. The abiotic part of the name describes readily recognizable, permanent physical features such as a soil type, a topographic feature, or a combination of the two. Some examples are sandy, silty, clay upland, saline, or meadow. Sites that have similar soil and topography may exhibit significant differences in their plant communities due to differences in micro climate. In such a situation, a biotic component is added to the name to differentiate among ecological sites occupying similar soil and topography but differing in species composition. By convention, the biotic name consists of one or two overstory dominant species and one or two understory dominant species. Subnames may be used if there are codominants in a community or the community has more than two layers of vegetation. Some examples are *Bouteloua chondrosioides/Mimosa dysocarpa* and *Mimosa dysocarpa–Cassia leptadenia/Bouteloua hirsuta*. The dominant species of the predominant life-form are listed first, followed by the dominant species of the next dominant life-form or understory species. The names of species of two different life-forms or layers are separated by a slash (/) and of the two or more codominants, by a dash (–). A common practice is to omit the abiotic component and to name ecological sites after dominant biotic characteristics.

7.1.2 Describing Vegetation Units

A good description clearly presents features that characterize a vegetation unit. It should not be heavily oriented toward a single use of the resource such as grazing, forestry, or recreation, but should cover all aspects of multiple resource use. Information for a complete vegetation description should include cover, density, and biomass for all major herbaceous species, and cover and density for shrubs and trees. When possible, common names should be listed along with scientific names of all plant species present. Two examples of typical ecological site descriptions from southern Arizona follow.

Site 1. Bouteloua chondrosioides/Bouteloua hirsuta/Mimosa biuncifera. This site is characterized by the presence of *Bouteloua chondrosioides* as the major species (20.3% cover). *Bouteloua hirsuta* and *Aristida ternipes* are the most common associates and are found occasionally as dominants in small, localized areas. Five of the six *Bouteloua* species found in the area occur in this ecological site. However, these species occur much less abundantly (<1.0%) than the two major *Bouteloua* species already mentioned. *Lycurus phleoides* occurs infrequently (2% cover), while *Aristida divaricata* occurs much more frequently (5% cover).

This site is easily observed in the field. The primary identification feature, aside from the vegetation, is that it occurs on flat topography. A major criteria for further delineation of the site is a slope of less than 5%.

The most abundant forbs are *Croton corymbulosus, Brayulinea densa, Evolvulus sericeus,* and *Sida procumbens.* However, none of these forbs exceed 2% cover values on this site. Annual forbs are quite abundant, as well as in other associations in the area. These forbs are easily identified but are also extremely variable in their occurrence from one year to the next. Therefore, individual information concerning these annuals is not used in the vegetation description.

The most abundant shrub is *Mimosa biuncifera,* with a cover value of only 2%. *Acacia angustissima,* along with *Haplopappus tenuisectus,* also occur in this vegetation association. Small amounts of *Desmanthus cooleyi* and *Haplopappus spinulosus* are observed on this site.

The standing crop values for the site, as expressed by the five life-form classes of vegetation, reveal that perennial forbs obtain their greatest production on this particular site. Perennial forbs are noted to have as much as 35 g/m^2, while the perennial grass component produces about 150 g/m^2. Annual forbs and annual grasses have values of 15 and 10 g/m^2, respectively. Shrubs are noted to produce 5 g/m^2, on the average, for this site.

Soils for this association are low in nitrates and have a 2% sand content. Iron content is rather high at 23 ppm. Phosphorus values are intermediate for this type. The pH of these soils averages 5.5.

Site 2. Eragrostis lehmanniana–Aristida divaricata/Bouteloua chondrosioides/Aristida ternipes. This site is characterized by the occurrence of *Eragrostis lehmanniana*, with an average cover value of 16%. This particular site has a codominant of *Aristida divaricata* with a cover value of 12%. Common grass associates include *Aristida ternipes*, *Bouteloua chondrosioides*, *Bouteloua gracilis*, and *Bouteloua hirsuta*. *Lycurus phleoides* occurs as a minor species.

The site can be easily identified by the striation pattern in the aerial photographs and exists because of reseeding of some areas. *Eragrostis lehmanniana* was planted during the 1950s, and the vegetation is slowly returning to its former state, that of site 1.

Common forbs for this association include *Brayulinea densa* and *Sida procumbens*, while *Convolvulus incanus* and *Evolvulus sericeus* are present in small amounts. The only shrublike species occurring in a significant amount is *Haplopappus spinulosus*, with an average cover value of approximately 1%.

Standing crop estimates obtained for the site indicate that the annual grasses produce the lowest with 5 g/m². Perennial grasses are noted to produce an abundance of biomass with a value of 200 g/m² on the average. Forbs are significantly lower in their contribution to the total biomass, with values for annual and perennial forbs being 10 and 15 g/m², respectively. Shrubs produce only 5 g/m², on the average, for the site.

Soils of this association are sandy with low nitrate values (0.80 ppm) and have a pH of 5.0. Phosphorus, potassium, and iron are found to be 10, 200, and 20 ppm, respectively. Organic matter is low in all soils and is only 1.0% in soils of the site.

7.2 BASIC CONSIDERATIONS

Vegetation monitoring and evaluation usually requires a detection of change in one or more characteristics of plant species over a time interval. No monitoring effort will succeed unless a basic inventory is made according to long-term needs of the program. That is, all information necessary for evaluation of changes will need to be identified according to species and specific measures needed. Monitoring of vegetation is not difficult if conducted properly. However, a plan for monitoring must be developed before a baseline inventory is made. Then, measures of vegetation characteristics will be comparable throughout the monitoring period. That is, one should know at the outset if plant cover is to be made on basal cover rather than aerial cover. Once characteristics are selected for measurement, a critical analysis should be made of the plan to determine any weaknesses likely to occur in monitoring efforts.

Specific effort should be made to measure plant characteristics which can be monitored to detect change. For example, plant cover should be determined in such a way that a significant change, even of small magnitude, is observed within a meaningful time interval. That is, if a change of one percent in a species cover value is considered to be ecologically significant, then obviously the measurement has to provide an estimate of cover in 1% increments. Then cover class intervals of widths greater than 1% are not acceptable. It follows that the level of precision desired should be determined at the outset of the measurement process. Plant density and biomass data of individual species, or collection by life-forms, must meet the same general principle as that followed for cover estimates.

To develop an understanding of need to select desirable precision levels prior to monitoring, again consider plant cover estimates. Other measurements may also have similarities to those of cover. Plant cover is often subject to a wide range of errors. This happens because estimation is often carried out visually rather than by objective methods. That is, plots are used and cover is estimated by some class interval other than 1% levels. If widths of these intervals are greater than 5%, then individual species cover values for most herbaceous plants will be overestimated unless such species are dominants. This is particularly true in semiarid and arid vegetation types where individual species, other than some dominant, seldom exceed a 10% cover value within a stand.

Additional problems occur in estimating plant cover, total or individually for species, when points are used to record hits. Basically, the same errors occur as those mentioned regarding class intervals. Compare the visually estimated cover with intervals of 10% to the estimate obtained from a set of 10 points in a frame. The 10 points provide an estimate of cover to the nearest 10%. Repetitions of the 10 point-frames and averaging does not provide closer estimates than 10%, since the basic unit of measure provides intervals of 10% (one hit out of ten = 10% cover). If a 1% precision level is desirable, then 100 points per unit must be used (one hit out of 100 = 1% cover). Other precision levels are also based on number of points used. For example, a 2% level of precision in measurement of cover requires 50 points, and so on. A change in percent of cover can only be detected at the level provided by the measurement process.

Problems in monitoring of biomass changes may be avoided by deciding initially which species or groups of species will be monitored. Obviously, if individual species biomass data is needed, then data are collected on individual species of interest, in which case plot size and shape should be optimum for species combinations. Other decisions include whether to separate total biomass of plants into components of leaf, stem, flowers, roots, and so forth. Moreover, consideration should be given to the use of destructive sampling; that is, harvesting or use of indirect techniques to estimate biomass changes.

Density measures are not necessarily easier to make for determination of change in vegetation. Definition of an individual is a problem. Clear descriptions of an ecological plant unit as an individual should be recorded and followed throughout the monitoring period. Otherwise, data may not be comparable to detect meaningful changes, if any, that may occur.

In summary, major considerations given before and during a monitoring and evaluation effort include:

1. Determine data use; for example, the monitoring of change in production of forage plants for wildlife. Then forage species must be identified initially.
2. Determine the measure needed. If measurement of cover is to be made, then will litter also be considered as to change over time? Selection of a set of measures to be made will determine the cost of the monitoring effort. Therefore, select the meaningful measures only.
3. Determine level of sensitivity to detect a change in vegetation. Changes in biomass of minor individual, but ecologically important species are detected only if data are collected for these individual species; the tendency is to collect such data for commonly occurring species in a monitoring program which may not provide significant results.
4. Determine the level of accuracy needed. Close estimates of density to actual values may be needed to detect a change in the abundance of a threatened species in contrast to a visual estimate of abundance of species for a general community description.
5. Select techniques to be used without change for the duration of the monitoring effort. Comparability of data can be made only if techniques provide comparable data. For instance, change in plot size and/or shape may alter comparability of data collected.

7.3   VEGETATION CONDITION AND CHANGE

The terms "range condition" and "range trend" emerged in range management literature after Sampson (1919) related the ecological principles of plant succession and climax to the condition of range vegetation and the direction of its trend. It soon became a tool for the management of grasslands and adjusting livestock numbers. Since then, extensive literature has been published on this subject because the methods involved are useful for evaluation and monitoring of vegetation changes, not just for livestock grazing purposes (Brown 1954). Grazing by domestic livestock is just one use of vegetation, and in many parts of the world, domestic and wild animals compete for grazing resources. Furthermore, harvesting of woody shrubs and trees for fuel compete with browse users. Therefore, wildlife and range managers are concerned with the physiological condition and species composition of vegetation.

Multidisciplinary interests in monitoring vegetation promotes the use of the terms "vegetation condition and change" rather than "range condition and trend." For example, it is well known that vegetation plays an important part in energy flow, and vegetation condition along with energy flow and utilization of native vegetation plant composition has a direct bearing on water yield of a watershed. Furthermore, demand is increasing for recreational use of natural areas, and park managers are concerned with the impact of human activities on vegetation. Finally, an interaction of biotic and abiotic factors on vegetation makes it more appropriate to refer to vegetation condition and change rather than to a singular use of vegetation by grazing animals.

The term "vegetation condition" relates the current status of vegetation to the potential of the site to produce a given vegetation unit. For instance, there is a general relationship between vegetation condition and stages in secondary plant succession. The more advanced the successional stage, the better the vegetation condition. However, from the standpoint of range management, the concept of range condition is one of practicality. A subclimax vegetation community may be more useful for livestock grazing than a climax community. Then, excellent range condition may be a subclimax vegetation rather than the climax vegetation preferred by those interested in pristine conditions for vegetation.

7.3.1 Vegetation Condition Analyses

Vegetation condition may be divided into as many classes or categories as needed. Classes or categories need only serve a meaningful interpretation. Several possibilities for describing the condition or stage of vegetation in relation to pristine vegetation are given in Table 7.1. Classes are based on the percentage of species composition found in climax vegetation. Then condition class or category is a descriptive term for the present condition of the vegetation of the area. Plan A of Table 7.1 may be useful for rangeland managers, and vegetation preservationists may prefer Plan B or C, since both latter plans require a higher percentage of the original vegetation to be present.

Vegetation condition classes were originally developed for rangelands

TABLE 7.1 Possible Plans and Classes for Vegetation Condition Classification

	Vegetation as Percent of Climax		
Condition Class	Plan A	Plan B	Plan C
Excellent	91–100	96–100	91–100
Good	76–90	76–95	71–90
Fair	51–75	51–75	56–70
Poor	26–50	36–50	46–55
Unacceptable	<25	<35	<45

and employed a measure of total herbaceous plant cover only. This single measure was found to be deficient since total plant cover could actually increase when grazed by livestock. That is, original species desired by large herbivores will decrease and are replaced by less desirable forage species. These latter species may provide an increase in total plant cover. Therefore, plant species composition was recognized as important in defining condition of vegetation. Then, growth form of grasses, such as bunch-types compared to single-stemmed types, does not overestimate the importance of species contributing to a condition class. Cover and biomass measures are made on a species basis, and either may be used to calculate species composition.

The best way to illustrate calculations of a given condition of a vegetation type from species composition is by way of an example. Table 7.2 provides information needed to calculate percentage composition from biomass measure by species from a range site. Note that the example includes a maximum allowable percent. This percentage is the maximum amount of the species found in the climax stage for the vegetation type or range site.

Dyksterhius (1949) selected descriptors of "increasers" for species that increase under grazing, "decreasers" as those that decrease, and "invad-

TABLE 7.2 Calculations to Determine Vegetation Condition Classification[a]

List of Plant Species	Weight (g)	Estimated Percent Composition	Maximum Allowable Percent Composition[b]	Percent Used
Blue grama (*Bouteloua gracilis*)	33	49	45	45
Western wheatgrass (*Agropyron smithii*)	7	10	5	5
Buffalograss (*Buchloe dactyloides*)	19	28	10	10
Sedges (*Carex* spp.)	3	4	7	4
Three-awn (*Aristida* spp.)	1	2	0	0
Squirrel tail (*Sitanion hystrix*)	1	2	0	0
Prickly pear (*Opuntia polycantha*)	2	3	2	2
Other forbs	1	2	1	1
Total	67	100	70	63[c]

[a]Data collected from a Loamy Plains site near Fort Collins, Colorado.
[b]Maximum allowable percent composition from a Loamy Plains site description at climax stage of development.
[c]63 = Good range condition, poor to fair ecological conditions.

ers" for those species that are not a part of the original vegetation composition but appear as grazing is increased. An ecological basis was used to arrive at an assignment of species into these descriptive categories. Essential to the assignment was a study of ungrazed areas consisting of similar vegetation.

Plant species respond differently to biotic and edaphic factors, and species will increase or decrease in numbers during the process of plant succession. When a climax or subclimax community undergoes retrogression, new species invade the area in addition to increases or decreases in the numbers of species already present. Vegetation condition classes should describe stages of secondary succession and should be regarded as being synonomous. Plants that increase under retrogression were called "increasers," those that decrease were "decreasers," and those that invade the areas were "invaders." Dyksterhuis (1948, 1949) used the terms "decreasers," "increasers," and "invaders" on the basis of ecological status of the species. Parker (1951) viewed the changes in vegetation from the perspective of livestock grazing and coined the terms "desirables," "intermediates," and "least desirables" in place of "decreasers," "increasers," and "invaders," respectively. The U.S. Soil Conservation Service (SCS) used the decreaser, increaser, and invader concepts that were based on responses to grazing, but these terms now refer to ecological responses to site characteristics. Most U.S. Forest Service regions use Parker's categories of desirable, intermediate, and least desirable categories. These categories have been questioned by some workers as to their relationship to the real status of plants in succession. Yet, the descriptions do allow for a relative ecological assessment to be made of a vegetation type, range site, or an ecological response unit. A relationship between intensity of defoliation, and the relative proportion of decreasers, increasers, and invaders for a hypothetical grassland site, is shown in Figure 7.1.

7.3.2 Vegetation Condition Change

A knowledge of the direction of secondary succession or vegetation condition change is needed to determine whether the vegetation condition is improving or deteriorating under the existing biotic or abiotic influences. Although objective determination of a change requires monitoring over a long period of time, there are some subjective methods for estimating the change at a given point in time. Reproduction of perennial plants, biomass accumulation, and amount of litter are important factors in the determination of vegetation change. These factors are considered by the USDA Soil Conservation Service and are listed on a score card. Each factor is given a rating of 0–4. A sample score card used by the U.S. Soil Conservation Service is given in Table 7.3. A similar card, containing needed additional information, can be constructed to detect changes in any vegetation.

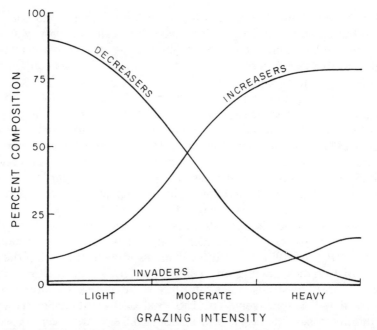

Figure 7.1 Relationship between intensity of grazing and percentage of decreasers, increasers, and invaders (Redrawn from Stoddart, et al. 1975, *Range Management.* Copyright © 1975 by McGraw-Hill. Used by permission).

7.4 MONITORING CONDITION AND CHANGE

Vegetation measurement most often used to monitor condition and trend are plant biomass, cover, density, and frequency, as mentioned earlier. In recent years, plant ecologists have expressed concern about the usefulness of these characteristics in monitoring vegetation condition and change. Advantages and disadvantages of which measure to use for monitoring purposes were discussed by Gardiner and Norton (1983), West (1983), and Winward and Martinez (1983). The reason for concern is seen in the behavior of attributes. For example, plant biomass is highly responsive to climatic fluctuations. Therefore, it has a tendency to follow climatic patterns, which may give a false impression of change. Furthermore, changes in proportions of biomass of all species do not respond in proportion to total precipitation, and its temporal distribution. Then, biomass changes may reflect changes because of management, grazing, or other perturbations controlled by humans.

Plant cover not only emphasizes the dominant species in a community but also gives greater emphasis to certain plants. A species may decrease in numbers, but its cover may actually increase significantly because the remaining plants become larger. Similarly, a few large plants of a species

TABLE 7.3   A Guide for Rating Range Trend

Considering the potential for this site, class of livestock, and season of use, rate the following by giving each item a rating of 4 to 0. Interpolate between definitions of 4 and 0 for ratings 3, 2, and 1.

Trend Indicators

Reproduction:
Decreasers producing abundant seedlings, various aged plants, and
  healthy tillers, rhizomes, stolons .... (4)
Decreasers and increasers not reproducing .... (3)_____
Plant residue and current utilization:
Decreaser litter is abundant for the site and current utilization of de-
  creasers is not destructive .... (4)
Decreaser, increaser, and invader litter not accumulating and range
  utilization is destructive .... (0)_____
Composition changes:
Evidence of strong increase in species most attractive to livestock, and
  decrease in species of low palatibility.............................. (4)
Evidence of strong decrease in species most attractive to livestock and
  increase in unpalatable increasers and invaders .... (0)_____
Plant vigor:
Decreasers are healthy, robust, not dying, well rooted .... (4)
Decreasers and increasers dying, shallow rooted .... (0)_____
Soil surface conditions:
No visible signs of accelerated erosion, past erosion being healed by
  decreasers or increasers, soil surface friable and not crusting........ (4)
Very obvious accelerated erosion, past erosion not being stabilized, soil
  surface crusting, not friable.. (0)_____
Total .... _____
Generally ratings totaling more than 12 indicate an *improving* trend—below 12
  indicate a *declining* trend.
Range Trend   Improving_____   Declining_____

From U.S. Soil Conservation Service, *National Range Handbook* 1964.

may die and be replaced by a large number of younger plants without any change in the cover value of the species (Fig. 7.2). In both these cases, shown in Figure 7.2, there is a definite trend in succession, but the cover indicates no change in the successional status of the community. Like biomass, cover is also greatly affected by climatic fluctuations. To some extent, this problem can be overcome by measuring basal cover rather than the foliar cover. Basal cover is a useful measure for monitoring tree species, bunch-type grasses, and sod-forming grasses. It cannot be used effectively to monitor rhizomatous grasses, prostrate plants, forbs, or multistemmed shrubs.

A

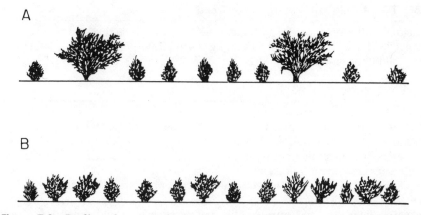

B

Figure 7.2   Profiles of two sagebrush steppe communities having similar brush and grass cover. Community A has only mature sagebrush; community B has only immature sagebrush (From N. E. West, 1983. "Choice of vegetation variables to monitor range condition and trend." Proc. Int. Conf. on Renewable Resource Inventories for Monitoring Changes and Trends. Copyright © 1983 by Oregon State University.).

Density may also be used as a measure for monitoring changes in vegetation. In particular, bunchgrasses, large forbs, shrubs, and trees can be easily counted in an appropriately sized plot. On the other hand, density of some species is inversely related to plant size, and small stature species will be emphasized in counts. Furthermore, there is difficulty in counting when plants have rhizomatous and stoloniferous reproductive forms. If an ecological individual can be determined, then counts are made accordingly. That is, define, as an individual, that part of a clone, mat, or other unit that will survive if a vertical separation of the unit occurred beneath the unit into the soil. Often, a simple tracing around the unit where bare ground occurs continuously will define an individual unit. If small annuals occur in large numbers, use plots with relatively small areas (0.1 m²). Otherwise, plots of sizes 1 m², 16 m², and 100 m² may be appropriate for herbs, shrubs, and trees, respectively.

Frequency may be used for monitoring vegetation change as it is one of the easiest vegetation characteristics to measure. Frequency is related to density as discussed in Chapters 4 and 5. Frequency is not an absolute measure and depends on the size of the quadrat used for measurement. Because frequency data are dependent on plot size and occur in nonrandom dispersion patterns, they cannot be used to calculate percentage species composition, as several workers have suggested. However, frequency data can be used to measure relative changes occurring by species and in their respective distributions over an area.

7.5 MEASUREMENT OF VEGETATION CHANGE

Choice of technique to measure vegetation change depends on answers to a number of questions. Is interest at species level, life-form, or for forage species only? Does the time interval needed include an annual growing season, several climatic cycles, or grazing intensities? Several approaches are given, one or more of which may be used, to monitor vegetation changes.

7.5.1 Individual Plant Approach

Individual plants of species can be monitored by use of permanently located plots. That is, an individual may be observed as to phenological and reproductive stages occurring over a season or years. Of course, all measurements must be non-destructive and employ methods given in previous chapters. Survival data, reproductive rates, and characteristics of cover, density, and biomass can all be measured for change.

Survival is a Bernoulli variable (values of 0 or 1) because a plant either lives or dies. However, as a series of observations over time, it can be modeled as a multinomial function and it is possible to estimate survival rates. The method of maximum likelihood can be used to obtain best estimates of survival rates and these estimates are then used to formulate a series of more complex models to explain variation in data. The models are tested by analysis of likelihood ratios, and the procedure allows testing of climatic, biotic, and environmental factors for their importance as casual factors in the process of change. An example to demonstrate this approach is given by Gardiner and Norton (1983).

7.5.2 Multi-Species Approach

No single vegetation characteristic is the best index of the vegetation condition. West (1983) recommends collection of as many types of vegetational data as possible to detect a significant successional shift or change in vegetation. Multivariate data analyses through mathematical, statistical, or graphic techniques would give more reliable results than a single variable. West (1983) used dominance–diversity graphs to combine variables for a better understanding of the vegetation condition and change. He plotted the natural logarithm of green biomass on the vertical axis and species abundance ranking on the horizontal axis (Fig. 7.3). The early seral stage (plot 2) was characterized by more species but also was less dominant by weight. In comparison, the late seral stage (plot 12) was dominated by trees with a sparse understory. Any change in the slope of the lines, as a result of sampling at some future point in time, indicates successional change.

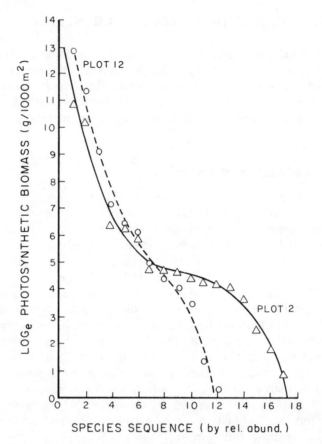

Figure 7.3 Dominance–diversity curves for two pinyon–juniper woodland communities of differing seral status. Plot 2 is early seral, and plot 12 is late seral (From N. E. West, 1983. "Choice of vegetation variables to monitor range condition and trend." Proc. Int. Conf. on Renewable Resource Inventories for Monitoring Changes and Trends. Copyright © 1983 by Oregon State University.).

7.5.3 Frequency Approach

As previously stated early in Chapter 4, frequency of a species depends on plant size, spatial distribution, and density. Therefore, vegetation condition classification, based on frequency sampling, is site specific. Mosley et al. (1986) employed multivariate statistical techniques to determine range condition from frequency data. A nested frame with 5 different plot sizes (5 × 5 cm, 10 × 10 cm, 25 × 25 cm, 25 × 50 cm, and 50 × 50 cm) was used for frequency sampling (Fig. 7.4). Data were collected on 18 sites in each of the three condition classes (good, fair, and poor) in mountain meadows of central Idaho.

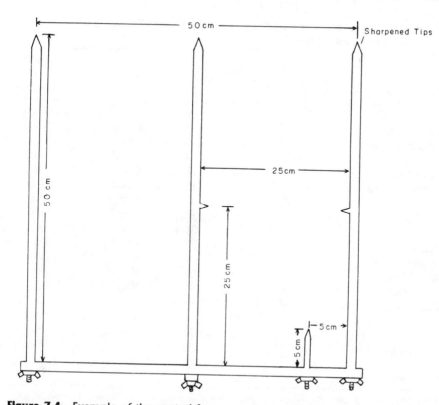

Figure 7.4 Example of the nested frequency frame used in Region IV, U.S. Forest Service trend studies (From USDA 1982).

Percent composition of each species was tabulated according to desirable, intermediate, or least desirable categories. Discriminant analysis procedures were used to classify sites into range condition classes, based on their relative frequency data for species. The analysis procedure yielded classification equations for each of the three vegetation condition classes. The equations were of the form

$$S_j = C_{j0} + C_{j1}Y_1 + C_{j2}Y_2 \tag{7.1}$$

where Y_1 and Y_2 are the percentages of desirable and intermediate species, respectively, and C_j are the weighting coefficients. Each site received a classification score S_j for each condition class, and the site was classified into the condition class for which it received the highest score. The data were first analyzed for a 10 × 10-cm plot, and then by continuous summation of data from each plot size. The summation of data did not result in classification improvement. The procedure resulted in correct classifi-

TABLE 7.4 Site Comparison of Range Condition Classification based on Summation Frequency Data

Site Name	Composition (%)[a]			Actual Condition Class	Classified by Frequency
	D[1]	I	L		
Cache Creek	67	18	15	Good	Good
Elk Meadow	57	35	8	Good	Good
Hartley Meadow	64	23	13	Good	Good
Poker Meadow	64	16	20	Good	Good
Sater Meadow	48	41	11	Good	Good
Stanfield Meadow	50	35	15	Good	Good
Bearskin Meadow	54	10	36	Fair	Fair
Corduroy Meadow (a)	48	35	17	Fair	Fair
Dead Cow Meadow	47	22	31	Fair	Fair
Pen Basin	57	16	27	Fair	Fair
Stanley Creek	38	31	31	Fair	Poor[b]
Ayers Meadow	34	15	51	Poor	Poor
Big Meadow	52	5	43	Poor	Poor
Bruce Meadow	49	18	33	Poor	Fair[b]
Corduroy Meadow (b)	37	42	21	Poor	Fair[b]
Little East Fork	39	29	32	Poor	Poor
Tyndall Meadow	45	19	36	Poor	Poor
Accuracy					83%

From Mosley et al. 1986. Determining range condition from frequency data. *J. Range Manage.* **39**: 564.
[a] D[1] = desirables, I = intermediates, L = least desirables.
[b] Misclassified sites.

cation of 15 out of 18 sites (Table 7.4). A condition guide (Fig. 7.5) was developed for use with relative frequency data from a 10×10-cm quadrat to estimate vegetation condition.

There is no single procedure for measuring vegetation change. The choice of an appropriate procedure depends, to a large extent, on the objectives of the study and the vegetation type. Generally speaking, multivariate methods provide more reliable results than univariate methods since several variables and/or several species are usually involved in monitoring.

7.6 SELECTION OF A MONITORING PROCEDURE

7.6.1 Vegetation Measurements

A well-conceived sampling design is a prerequisite to any monitoring program. The choice of a sampling design is dictated, in part, by the objectives, the vegetation type, the vegetation characteristics to be measured, and the availability of financial and technical resources.

Designs most frequently used are random sampling and permanent plots. Vegetation sampling, for both inventory and monitoring, is conducted by vegetation types or subtypes. Sampling for monitoring is, however, usually restricted to selected locations within a vegetation unit. Sampling locations within a vegetation unit may be selected at random at each measurement time, or repeated measurements may be obtained on the points selected at random the first time. The position of selected points is permanently marked on the ground by driving steel rods to facilitate exact relocation of the plots or lines. Areas selected for monitoring studies are called "indicator areas" or "key areas." These areas of vegetation should

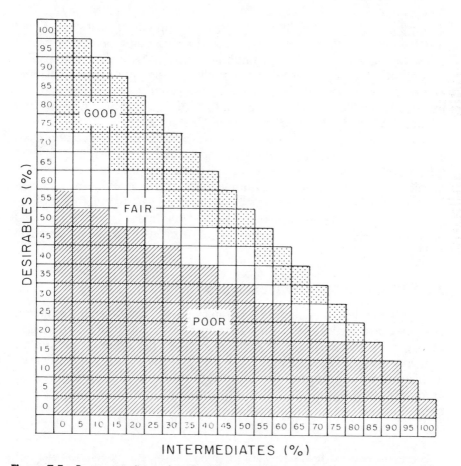

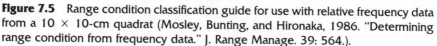

Figure 7.5 Range condition classification guide for use with relative frequency data from a 10 × 10-cm quadrat (Mosley, Bunting, and Hironaka, 1986. "Determining range condition from frequency data." J. Range Manage. 39: 564.).

be those that will respond the quickest to a perturbation. Obviously, key areas or indicator areas will vary according to a given perturbation.

Permanent plots are not suitable for destructive types of biomass measurement. The problem can be resolved by use of double sampling or other indirect sampling procedures discussed in Chapter 6. Best estimates of foliar or basal cover, for the purposes of monitoring, are obtained by measurement from permanently located plots or lines. Permanent plots can be two-dimensional (square, rectangular, or circular) or one-dimensional (line transect). Cover on two-dimensional plots may be monitored by charting, photographic, or other techniques. Cover sampling along a line transect is conducted best by point sampling or recording the length of line that intercepts a plant or plant part.

Frequency sampling may be conducted by random sampling at each time or along permanently located line transects. The plots may be randomly located along the line transect at each point of observation, but usually these are systematically located. The step-point method or nested plots may be used for random sampling. Frequency sampling along a line transect may be conducted by point sampling, nested plots, or a small loop. The point-sampling data for cover sampling can also be used to calculate frequencies of the species. The procedures for estimating cover and frequency have been described in detail in Chapter 4.

As previously noted, density is one of the least used vegetation characteristics for monitoring vegetation trends. This is obviously the result of time needed to count individuals of a species in a bounded area. However, density can be estimated by counting the individuals of species on permanently marked two-dimensional plots or by repeated random sampling. Density may also be estimated by plotless methods. Both the plot and plotless methods of density sampling have been discussed in detail in Chapter 5.

7.6.2 Spectral Imagery Techniques

Monitoring and evaluation of vegetation characteristics on a time sequence basis may be accomplished by use of imagery available from ground-based photography and electronic spectral imagery or from airborne and satellite imagery. A video system mounted on low-flying aircraft is now providing multispectral imagery also. This method is readily available, and imagery processing is inexpensive compared to that obtained from satellite platforms. Images from a video camera can be viewed immediately by playback, and resolution is as small as one meter. Digitizers are available for processing film.

Precision needed to detect certain small changes in vegetation may be difficult to obtain for several reasons. If United States satellite imagery is used, then a unit of measure ranges from 0.1 to 1.4 ha in size, depending on the sensor and band used. This compares to a French satellite system providing a resolution of 0.01 to 0.04 ha in size. Repeat coverage time of satellite systems vary from 16 to 18 days. Obviously, any vegetation change has to be sufficient to be detected over an area larger than that detected by the system. Additional factors affecting precision of any image obtained are altitude differences, atmospheric distortions, and topographic position of the observed site.

Satellite imagery has been successful in determining vegetation condition and changes occurring over time. For example, Lacaze et al. (1983) studied rangelands in Tunis and applied imagery obtained from LANDSAT to monitor changes in vegetation characteristics.

7.6.3 Sequential Photographs

Photography is valuable for evaluating vegetation changes and has been used since the late 1800s (Hocker 1978). Photographs can be made over time of the same area as needed. Qualitative characters (abundant, small, large, etc.) and quantitative characters (density and cover) can be determined from photographs containing appropriate resolution. In fact, photos are better in detecting certain changes before conventional methods can do so (Gardiner and Norton 1983, West 1983). Another favorable aspect of high resolution-repeated photography is the possibility of changing the characteristics of vegetation to be measured. If a measure is needed later, then the record is available.

Detailed instructions on the use of time-sequence photographs to study vegetation changes are provided by Rogers et al. (1983) and Owens et al. (1985). Exact positioning of the camera at each time for each plot is the major consideration. Permanent markers can be used to accomplish this. Then, vertical and horizontal measures are used to position the camera. Flash units can be used to eliminate sunlight differences at varying times of the day. If possible, stereophotographs should be obtained so that more detailed and accurate measurements can be made, as described in Chapter 4.

Objectives of a vegetation monitoring and evaluation program will determine the size of area to be photographed, as well as the time intervals between photographs. Plot sizes range from 0.1 m to 55 m^2 for grasslands to shrublands, respectively. The larger the plot, the higher the location of the camera. Then, a boom may have to be used in place of a tripod (Goodwin and Walker 1972, Owens et al. 1985). To relocate exact camera positioning over time, rods can be driven into the ground at each plot. Three such rods over which a traditional tripod or a boom can be positioned are preferable. A plumb bob hung from camera center to a rod driven into the ground at the plots' center can also assist in relocation for repeated photography.

7.7 HERBAGE REMOVAL

Historically, populations of herbivores have followed cycles of available forage. Increases in number of herbivores result in increased utilization of plants and, conversely, scarcity of forage results in mortality and/or the migration of animals. In addition, natural catastrophies such as drought, floods, or fires may result in large-scale destruction or alteration of vegetation. However, natural processes always lead to a recovery of vegetation systems, and the rate of such recovery depends on severity of the change. Then, increases in the availability of forage or habitat again result in a subsequent increase in the number of animals. These cause–effect cycles differ in length, but are the essence of natural ecosystems. These cycles are also modified by humans and their domestic livestock.

Modern humans have interfered with natural processes of ecosystems. Pastoralism is one of the earliest professions, and people traditionally moved their livestock from place to place in search of forage for animals. This migratory lifestyle helped vegetation to periodically recover from heavy utilization by domesticated animals (Brown, 1954). However, increases in human population and a corresponding increase in livestock numbers have permanently altered natural vegetation in most areas of the world. In the Old World, pastoralists have severely depleted grazing resources over a period of centuries. On the other hand, destruction of grazing areas in the new world has also occurred recently.

Grasslands are often found in combination with environmental factors that cause fragile ecosystems to exist that are prone to easy disturbance by changes in biotic factors. A quantification of herbivory and its impact on these ecosystems are essential; not only for maximum sustained production of livestock and livestock products but also for other environmental considerations. Land managers and others are faced with the problem of determining the current status of natural vegetation resources, the degree to which these can be exploited, and subsequent effects of various levels of exploitation.

Much of the land area of Earth is used primarily for grazing by domestic livestock. Wildlife, rodents, and insects are an integral, but less visible, part of these grazing lands. Yet, these latter herbivores not only consume large amounts of vegetation but also cause significant change to vegetation at times. A lack of an intimate knowledge of the dynamics of herbage removal and its impact on the ecosystem has not only resulted in deterioration of vegetation condition but has also accelerated the processes of desertification in many parts of the world. Attention is needed for vegetation measurements to estimate the amount of forage removed by herbivory and to monitor vegetation, especially for changes over time.

7.7.1 Carrying Capacity

The amount of biomass or herbage production on a vegetation community at a particular point in time can be estimated using one or more of the methods described in Chapter 6. However, estimates of biomass alone are not sufficient to determine how many herbivores can graze over a given time interval to obtain maximum sustained production of herbivore biomass consistent with other land uses. Generally, grasslands occur in arid and semiarid regions where weather is extremely variable. However, forage production is a function of not only total precipitation but also its seasonal distribution. Any decision, then, to graze certain numbers of animals, based on minimum forage production in any one year, will result in underutilization of forage resources. On the other hand, a sliding scale of stocking grasslands, on the basis of the amount of forage available each year, is not feasible, either, from the management standpoint. Therefore, long-term averages are used to graze lands, and monitoring on a 5–10-year interval is acceptable in order to determine needed adjustments in animal numbers.

There are four important reasons why biomass estimates alone cannot be used to make a decision on the stocking rate of animals:

1. Not all herbaceous biomass is forage for a specific herbivore. Each class of herbivore differs in its consumption of biomass and according to herbs, shrubs, and trees.
2. Annual production of biomass for a plant varies according to species and grazing intensity effects.
3. Historical effects on biomass of plants is reflected during a given year of sampling. Therefore, utilization by herbivores may have to be reduced to allow plants to recover.
4. No technique used in determining plant biomass even closely resembles herbivory. So, estimates of biomass available for foraging is only approximate.

Some of the differences between grazing and harvesting are: (1) vegetation is harvested at a uniform height while animals graze at a convenient height, (2) natural preferences of animals cannot be simulated by a harvesting technique, and (3) a special effort must be made to determine the forage lost through trampling by grazing animals (Culley et al. 1933).

Grazing or carrying capacity is the point of equilibrium that exists between plants and animals. It is further defined as "the maximum animal numbers which can graze each year on a given area of grassland for a specific number of days without inducing a downward trend in forage production, forage quality, or soil" (Stoddart et al. 1975). Note that those interested in vegetation resources as livestock forage use a restricted definition for carrying capacity. The definition of grazing capacity varies slightly from author to author, but the basic consideration that many herbivores are utilizing these vegetation resources should always be kept in mind. Vegetation changes have occurred because of insects. Yet, the deterioration in vegetation condition was attributed to livestock grazing.

The first attempts to determine the grazing capacity for livestock were made in the beginning of the twentieth century. Some of the methods used are no longer employed by grassland ecologists but are briefly mentioned here because present-day procedures have evolved from these, and some should be used to evaluate and monitor vegetation in the present time. These are (1) the reconnaissance or the forage hectare method, (2) the point-observation-plot method, and (3) methods currently being used called "weight methods."

True carrying capacity for large herbivores can be determined only by stocking an area in acceptable condition with animals and monitoring the vegetation trend. If the stocking results in vegetation improvement, animals are too few and the number should be adjusted upward. If the stocking results in vegetation deterioration, animals are too numerous and the numbers should be adjusted downward. In all cases, herbivory by all other sources must be determined to adequately evaluate forage removed.

7.7.2 Reconnaissance or Forage Hectare Method

A reconnaissance method was developed between 1907 and 1910 by James T. Jardine, then Inspector of Grazing, U.S. Forest Service (Pickford 1940). The concept is still useful for practical purposes in grazing management and, with little modification, can be used for other purposes such as wildlife herbivory. Vegetation types and subtypes are mapped onto survey maps, and other features important to range management are also mapped and described (Brown 1954). The method defines a vegetation type as an area of vegetation recognizable by its general aspect, such as grassland and shrubland. A subtype is defined as an area within a vegetation type that is differentiated from other areas by floristic composition, density of vegetation, or plant vigor. The latter term is the general health appearance of plants and is measured indirectly from measures of plant size, reproduction, and so forth.

The procedure for estimating grazing capacity involves estimation of the average cover of total vegetation and the relative cover of component species. This is accomplished in a reconnaissance survey of the vegetation subtype. Portions of shrubs or other plants growing beyond the reach of animals are ignored for cover estimates. In two-layered vegetation (e.g., grasses growing under shrubs), the cover of each layer is estimated separately if both are available to animals. After a reconnaissance survey of the vegetation, a small bounded area is selected that represents the subtype. The cover and composition of the subtype are then visually evaluated. If important changes occur in the bounded area, the original estimates are revised. If differences exist between the subtype estimates of cover and those from the small area, then the estimates for the subtype should be re-evaluated. If the differences are of magnitude, then another small area should be selected for cover estimation, and the process should be continued until the average cover by species are in agreement with those estimated for the subtype. However, floristic composition differences may be found among small areas, so the average of several small areas should be substituted for those of the original estimates made for the subtype.

Prior knowledge of "proper-use factors" of each species is required to compute grazing capacity. Proper-use factors are percentage of current herbage production of species that are utilized when the vegetation is properly grazed. The proper-use factor of a species does not exceed the use it can withstand each year without loss of vigor, and experience allows such estimates to be made. Computation of grazing capacity by this method is shown in Table 7.5. Metabolic weight information, provided later, allows this method to be converted into units of use for any herbivore (see Section 7.7.4).

TABLE 7.5 Computation for Grazing Capacity by the Reconnaissance Method with a Cover of 21% on 332 Hectares[a]

Species	Percent Composition	Proper-Use Factor	Weighted Percent Proper Use
Bluebunch wheatgrass (*Agropyron spicatum*)	0.23	0.8	0.18
Idaho fescue (*Festuca idahoensis*)	0.09	0.7	0.06
Sheep fescue (*Festuca ovina*)	0.12	0.5	0.06
Common geranium (*Geranium* spp.)	0.03	0.6	0.02
Bitterbrush (*Purshia tridentata*)	0.08	0.3	0.02
Total	0.55		0.34

[a]Forage/hectare requirement = 0.5 forage hectares/cow month:
Forage factor   = cover × weighted proper-use factor
                    = (.21)(.34)
                    = .07
Forage/hectare   = forage factor × area
                    = (.07)(332)
                    = 23 hectares
Grazing capacity   = forage hectares × forage hectare requirement
                    = (23)(1.27)
                    = cow months (AUMs)
    29 or (29)(4)   = 116 deer months (see Table 7.6)

Proper-use factors for each species, and forage hectare requirements, are developed from long-term grazing records. The average number of animals grazed and the average grazing period are determined and reduced to an expression of cow months, sheep months, or other animal unit months (AUMs). For example, if 100 sheep spend 6 months on a subtype with 760 forage hectares, the forage hectare requirement would be 1.27 forage hectares per sheep month ($760 \div 100 \times 6 = 1.27$). It is presumed that the forage hectare requirement will be the same on similar areas. Grazing capacity calculations by the forage hectare method can be summarized as follows:

$$\text{Grazing capacity} = \frac{\text{cover} \times \text{proper use rating} \times \text{surface area}}{\text{forage hectare requirement}} \quad (7.2)$$

Defoliation studies on individual species have been conducted, and literature on specific species can be used to find reported effects of herbivores removing foliage. Such effects are used to determine proper-use factors.

7.7.3   Point-Observation-Plot Method

This method was developed and tested by Stewart and Hutchins (1936). The method is similar to the reconnaissance method except that cover is estimated in quadrats rather than over the entire area. Quadrats originally were circular in shape and 10 m² in area. Three to six such quadrats are estimated in subtypes smaller than 10 ha, 5–10 quadrats in 10–30 ha subtypes, and 10–20 quadrats in 30–250 ha subtypes. In "strip survey" methods, 20 quadrats per 250 ha are measured if the large ones are used. Quadrats are located systematically within the subtype. The basic unit of estimation is a 0.1-m² plot, described in Chapters 2 and 4, which is completely covered by foliage when viewed directly from above. Cover estimates require a period of training prior to sampling and involves mentally placing plants in a 0.1-m² frame in such a way that they constitute 100% cover. Cover is estimated by species. Grazing capacity is then computed the same way as in the forage hectare or reconnaissance methods. Sample computations of grazing capacity by the point-observation-plot method are shown in Table 7.6.

Neither the reconnaissance nor the point-observation-plot method will give exact estimates of grazing capacity over large regions because (1) production of forage is a function of volume of plants rather than their cover values, (2) proper-use factors of plants vary from place to place, (3) forage hectare requirements also vary from place to place, and (4) it is difficult to keep long-term records to determine forage hectare requirements of different kinds of herbivores. However, these methods do provide baseline estimates, and proper monitoring will provide information to be used

TABLE 7.6   Computation for Grazing Capacity by Point-Observation Plot Method[a]

Species	Percent Cover/ Quadrat Number 1	2	3	Total Cover (%) (A)	Average Cover (%) (B)	Proper Use Factor (C)	Weighted Forage Factor (B × C ÷ 100)
Blue grama (*Bouteloua gracilis*)	43	37	46	126	42	0.80	0.34
Western wheatgrass (*Agropyron smithii*)	11	7	13	31	10	0.70	0.07
Buffalograss (*Buchloe dactyloides*)	23	14	30	67	22	0.75	0.16
Sedges (*Carex* spp.)	3	6	4	13	4	0.50	0.02
Three awn (*Aristida* spp.)	5	2	—	7	2	0.50	0.02
Prickly pear (*Opuntia* spp.)							
Total	85	66	93	244	80	—	0.61

[a]Data collected from a Loamy Plains site near Fort Collins, Colorado.

for adjustments needed, if any, in herbivore numbers. Some of these problems have been overcome by the use of weight estimates of herbage production.

7.7.4   Weight Estimate Methods

Techniques of estimating biomass have been described in detail in Chapter 6. A rule of thumb often used is that if vegetation is in good condition, then 50% of forage can be utilized without any adverse effect on the health of the plants. Like most "rules of thumb," this suffers for lack of a quantitative basis. If the vegetation is not in good health, then proportionately more forage should be left ungrazed to enable the vegetation to recover physiologically. On the other hand, some species may be harvested at an 80% level and not be harmed. Once a decision has been made on the percent vegetation that can be removed by all known herbivore sources, and if the daily forage requirements of different types of animal are known, then the grazing capacity can be computed for different vegetation subtypes. Grazing capacity is normally expressed in terms of AUM, and a domestic beef cow is taken as a standard. An AUM is the amount of dry forage consumed by a cow and calf in 1 month (30 days). It is estimated that a cow and calf need 10–13 kg (24–31 lb) air-dry forage per day. Therefore, one AUM is equal, approximately, to 400 kg. The animal unit equiv-

TABLE 7.7 Approximate Number of Individuals per Animal Unit Based on Ratios of Metabolic Weights[a] for Mature Animals

Species	Approximate Weight		$x^{0.75}$	Ratio 98/x	Number per Animal Unit
	lb	kg (x)			
Cape buffalo	1200	545	112	0.87	0.9
Bison, cow, eland, horse	1000	455	98	1.00	1.0
Elk, zebra	600	272	67	1.46	1.5
Waterbuck, wildebeest	400	182	50	1.96	2.0
Hartebeest, topi	300	136	40	2.45	2.5
Mule deer	150	68	24	4.08	4.0
Sheep, impala	120	55	20	4.98	5.0
Pronghorn antelope, goat	100	45	17	5.76	6.0
Thomson's gazelle	50	23	10	9.80	10.0
Dikdik	12	5	3	32.67	33.0
Black-tailed jackrabbit	5	2.3	2	49.00	49.0

Reprinted from Heady, H. F. 1975, *Rangeland Management* p. 149.
[a] Weight 0.75 kg.

alents of other animals are given by the ratio of the metabolic weight of a cow to the metabolic weight of the other animal. The metabolic weight of an animal is three-fourths the power of its body weight in kilograms (Kleiber 1961). Animal unit equivalents of a number of species are given in Table 7.7. Sample grazing capacity calculations by the weight method are shown in Table 7.8. Note that decimal places are not used because g/m^2 weight are seldom obtained to the nearest 0.1 g. Once again, different vegetation types are used by various herbivores in addition to livestock. Some methods to determine such use are presented in later sections.

7.8 UTILIZATION

Grazing capacity estimates for vegetation types, ecological sites, or range sites are used only as a guide to determine stocking rates for livestock or wildlife. It is important to know at the end of the grazing season how much forage was utilized by the animals throughout the grazing season so that adjustments can be made if needed. Utilization is defined as the percentage of the annual production of forage that has been removed by the animals throughout a grazing period or grazing season. While the

TABLE 7.8 Computation of Grazing Capacity by Weight Method (Colorado Shortgrass Site)[a]

List of species	Clipped Green Weight (g/m²)					Average Green Weight (g/m²) (A)	Dry: Green Weight Ratio (B)	Average Dry Weight (g/m²) (A × B)
	Quadrat Number							
	1	2	3	4	5			
Blue grama	38	19	42	24	40	33	.62	20
Western wheatgrass	11	12	2	14	4	9	.58	10
Buffalograss	2	17	13	21	14	17	.60	10
Sedges	8	3	6	—	7	5	.51	3
Threeawn	2	5	3	1	—	2	.64	1
Forbs	10	13	9	12	15	12	.41	5
Total	71	69	75	72	80	78		49

[a]Species same as in Table 7.6. Vegetation subtype, Loamy Plains; vegetation condition, good; area, 750 ha; percent of biomass that can be utilized, 60%.

$$\text{Biomass per hectare (kg)} = \frac{49 \ (\text{g/m}^2) \times 10,000}{1000} = 490$$

$$\text{Grazing capacity (AUM)} = \frac{\text{biomass (kg/ha)} \times \text{area (ha)} \times \text{utilizable proportion}}{400 \text{ kg}}$$

$$= \frac{490 \times 750 \times 0.60}{400}$$

$$= 551 \text{ AUMs}$$

interpretation and use of the term "utilization" differs among authors, more commonly, utilization refers to (1) the use of total vegetation on a given area and (2) the use made of individual plant species. Stoddart et al. (1975) consider the second usage of the term "utilization" more appropriate than the first. The second definition is a better one than the first if utilization by individual class of herbivore is to be measured.

7.8.1 Ocular Estimate

The ocular estimate is a rapid and easy method for determining utilization percentage for vegetation over large areas. This procedure is somewhat similar to that of ocular estimates of biomass production described in Chapter 6. The only difference is that in a sampling unit, an estimate is made of the herbage removed in a sampling unit rather than of the biomass of plants in a sampling unit. The ability to estimate utilization is developed through training. That is, in an ungrazed area, sampling units are clipped to simulate different intensities of grazing, and an individual worker estimates percent utilization and compares it to actual utilization. Clipped herbage x is weighed; remaining herbage y is then clipped and weighed separately. Actual utilization is then the fraction of total biomass removed, expressed as a percentage:

$$\text{Percent utilization} = \frac{\text{biomass removed } x}{\text{total biomass } (x + y)} \times 100 \qquad (7.3)$$

Most technicians can be trained to estimate varying degrees of utilization within reasonable limits of accuracy. During the course of the survey, technicians should continue to test their ability to estimate utilization by clipping and weighing a few sample plots at the beginning of each day. If necessary, estimates may be checked again during the day. The number of samples required to estimate utilization within given limits of accuracy should be calculated using the method described in Chapter 3. Generally, 25–30 observations should be sufficient for estimating utilization in homogeneous native vegetation. The method is rapid and permits ample replication for statistical analysis of data.

A refinement of the method involves making estimates on individual plants instead of estimating total herbage removed from a sampling unit. However, this procedure is not as rapid as estimates by total vegetation removed from a plot, but there is less personal error since each observation is confined to a single plant. The individual plant procedure has a high correlation to actual weight removed and is, therefore, more suitable for detailed descriptions of utilization patterns. An essential prerequisite for both procedures, ocular estimate by plot and ocular estimate by plant, is that reference areas that are ungrazed must be set aside throughout the vegetation area. Reference areas are used for initial training and daily checking of the precision of ocular estimates. Ocular estimates are subject to all the errors of the weight-guess method. Accuracy can be improved by adjusting estimates by the double sampling method described in Chapter 6. That is

$$y = a + bx \qquad (7.4)$$

where y is the predicted amount of biomass removed as estimated visually and x is the amount found by clipping that same plot. The plot is first clipped of a given amount of biomass, then an estimate is made as to how much was removed. The clipped amount is then weighed.

7.8.2 Clipping and Weighing

This method involves clipping and weighing grazed and ungrazed plots either for total herbage or by individual species. Included are any species that provide herbage as forage. Therefore, trees, shrubs, and herbs are to be considered. If an area is grazed during the growing season, then one set of sample plots must be protected from grazing. At the end of the grazing season, measurements of biomass are taken on both plots protected from grazing and plots in the grazed area. The difference in biomass between two sets of data provides an estimate of utilization. If grazing occurs only during the dormant season, the procedure for determining utilization is slightly different. In this case, biomass measurements are made on the same area before and after grazing.

The paired-plot method requires establishment of pairs of plots or transects at randomly selected locations or key areas. One member of the pair is enclosed in a cage or a small fenced exclosure for shrubs and trees to protect it from grazing. This is referred to as the "cage comparison method." Plots are clipped at the end of the grazing season. Biomass from plots protected from grazing represents total herbage production, and biomass from grazed plots represents the unused herbage biomass. The difference between the two weights is the amount utilized, which is expressed as percent of total herbage production. Utilization may be determined for total biomass or by species. In the latter case, plots are clipped by species. One problem in determining utilization during the growing season is that the growth rate of species grazed is different from that of protected plants (Cook and Stoddart 1953).

Plots are sometimes located randomly both before and after grazing. Experience has shown that randomization of plots, both before and after grazing, does not give satisfactory results. Occasionally, biomass is greater after grazing than before grazing. Animals do not uniformly graze an area. Therefore, the plots should be placed in pairs at each randomly selected sampling points. One plot is clipped before grazing and the other after grazing. The difference in the over-dry weights of biomass before and after grazing gives an estimate of the utilization and is expressed as percent of biomass before grazing.

7.8.3 Reference Unit Method

Cassady (1941) developed a technique of estimating utilization that was plotless. In this procedure, a specific plant unit, such as a stem, leaf, twig, or a whole plant, is selected as the reference unit. Reference units will differ for different species of life-forms. A predetermined number of such units is clipped and weighed before the area is grazed. At least 25–30 units should be randomly selected, clipped, and weighed. Sampling points are selected at random or systematically, starting from a randomly selected point along one or more transect lines throughout the area. After grazing has ended, a number of similar units are again clipped and weighed. Utilization is the difference between weights of units clipped before and after grazing expressed as a percentage.

The selection of reference units (stems, leaves, or twigs) is purely subjective and can introduce considerable bias in sampling because the unit selected is considered to be that grazed by animals. This may not be so.

Another disadvantage of this procedure is that plant growth during the period between two sampling dates is not taken into consideration. The method, however, can be used effectively where animals remain on one location only for a few days and then move on. Therefore, growth before and after grazing is normally negligible. Otherwise, the procedure should be repeated at short time intervals (1 or 2 weeks) over the period of grazing. Then, difference methods are used to estimate total biomass production by the unit in the same way caged plots are used (Section 7.11) over a season sampled in several time units.

7.8.4 Stem Count Method

Percent utilization is a direct function of the number of stems on plants that are grazed (Stoddart 1935). A method for determining utilization is to count the number of stems grazed and the number ungrazed in a sampling unit. A sampling unit may be a line transect or a plot. If a line transect, then intercepted plants are noted to be grazed or not. In plots, counts are made of grazed and ungrazed plants. Large plots, then, will be time-consuming, and errors occur because it is difficult to keep an accurate tally of plants. The difference of total stem count minus the number of grazed stems, expressed as a percent of total number of stems, is an estimate of utilization. The only decision one has to make is whether a plant has been grazed or not. Pechanec (1936) tested the stem count method and found that the method did not compare well with other methods. The comparison is shown in Table 7.9. Large values obtained through the stem count method are attributed to the fact that stems are grazed at different height levels. As mentioned before, stem height and volume of biomass are not linearly related. However, this method is useful for monitoring utilization in a relative manner and is comparable from year to year for the same species on a site.

TABLE 7.9 Comparison of Stem Count Method with Other Methods of Determining Utilization

Method	Utilization (%)
Clipped weight	41
Stem count	70
Height	42
Ocular estimate by plot	
Observer 1	39
Observer 2	43
Observer 3	42

From "Stem-count method" by J. F. Pechanec, *Ecology*, 1936, **17**, 330. Copyright © 1936 by The Ecological Society of America. Reprinted by permission.

7.8.5   Height–Weight Methods

The height–weight method is based on the assumption that in some plants, weight and volume are correlated and volume can be used to predict weight (Lommasson and Jensen 1938, Crafts 1938). The method consists of determining percentage reduction in height due to grazing, and converting this reduction in height to reduction in volume. In fact, the terms "volume" and "weight" have been used interchangeably in literature, although this represents a misuse of terms. In the early stages of the development of this method, it was assumed that there was a direct relationship between height and weight of grasses. For instance, a species with an average height of 20 cm, grazed to a stubble height of 5 cm, was considered 75% [15 out of 20 (15/20)] utilized. However, it was soon realized that distribution of weight with height in plants was not a linear function. The relationship between the height of two grasses and volume of plant material is shown in Figure 7.6. This figure clearly indicates that the bulk of the weight of a grass plant is in the basal portion of some grass species and mostly close to the ground in other grasses.

Growth forms of different species are different; therefore, height–weight relationships (Fig. 7.7) must be prepared for each species before this method can be used. The picture of each species is actually taken to the field for the survey. Data may also be plotted as height–weight curves (Fig. 7.8).

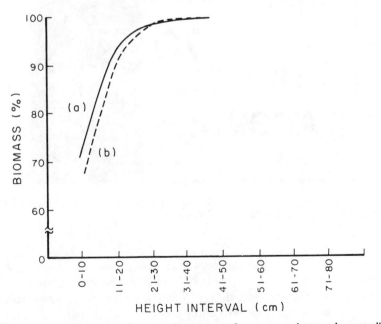

Figure 7.6 Percentage distribution of biomass of grass *a* and grass *b* according to height.

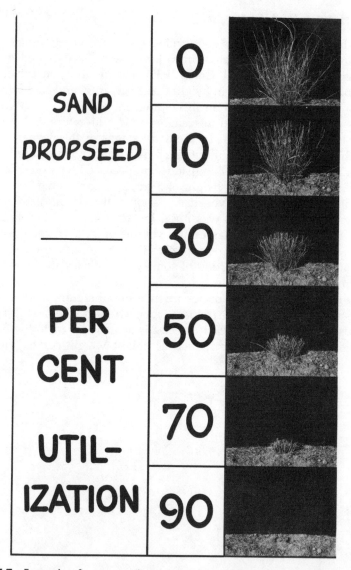

Figure 7.7 Example of a grass utilized according to various percentages of forage removed (courtesy of E. Schmutz).

The height–weight method assumes that growth form is reasonably constant for a species. However, growth form of herbaceous plants changes with changes in edaphic and climatic factors. Grass plants approximate a cylinder in years of good growth and a conical form in years when growth is slow (Clark 1945). Some species show greater variation from site to site, and some from year to year. However, in other species, growth form

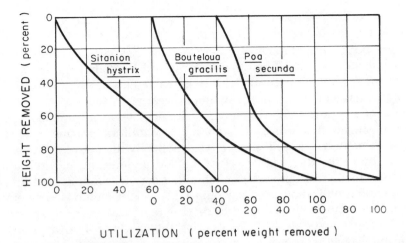

Figure 7.8 Three types of height–weight curve of range grasses, all with seed stalks: bottle-brush squirreltail (*Sitanion hystrix*) from Utah; blue grama (*Bouteloua gracilis*) from Colorado; and sandberg bluegrass (*Poa secunda*) from Utah (R. S. Campbell, 1943. "Progress in utilization standards for western ranges," J. Wash. Acad. Sci. 33.).

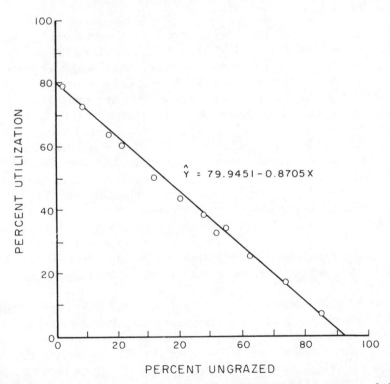

$$\hat{Y} = 79.9451 - 0.8705\,X$$

PERCENT UNGRAZED

Figure 7.9 Regression line showing the relationship between percent of all ungrazed important grasses on the Santa Rita Experimental Range and percent utilization (From M. E. Roach, 1950. "Estimating perennial grass utilization," J. Range Manage. 3: 183.).

297

changes both with site and weather. Weather has been shown to affect growth form more than any other factor (Caird 1945). In contrast, Heady (1950), in his work on the height–weight relationship of plants, found greater variation among sites than among years.

Reid and Pickford (1941) compared the height–weight method to the weight estimate method. Under conditions of uniform grazing by sheep, the two methods yielded similar results. But, with uneven grazing of grasses by cattle, the height–weight method underestimated utilization. The latter method is time-consuming, and it may take 2 or 3 times longer to determine utilization than it would with weight methods. Therefore, this method should not be used to determine utilization over vast areas. However, it may be used for monitoring the effects of utilization on experimental plots where destructive sampling, such as weight estimates, may not be desirable.

7.8.6   Plant Count Method

The plant count method combines the technique of stem count and height–weight or ocular estimate methods (Roach 1950, Hurd and Kissinger 1953). A regression equation is developed between percentage X of grazed or ungrazed plants and percent utilization Y. Pace transects are used, and the plant nearest to the toe is classified as grazed or ungrazed. The estimate for percent utilization of grazed plants is determined from the height–weight relationship. A regression equation is shown between percent plants grazed and percent utilization in Figures 7.9 and 7.10. There is a high correlation between the percent grazed plants and percentage herbage removal. However, note that Figure 7.9 is based on ungrazed plants, while Figure 7.10 uses grazed plants to determine utilization. Note further that one is linear, while the other is quadratic in nature.

Regression relationships of this type are usually site specific. That is, these cannot be applied any place except the site where the equation is developed. At high levels of utilization (say, greater than 70% of plants grazed), the method loses its sensitivity. For instance, Springfield and Peterson (1964) found that when 100% of some grasses had been grazed, the estimated utilization by this method was only 35%. Gierisch (1967) partially solved this problem for estimating utilization by counting only those plants that had been utilized more than 30%. He found that if thurber fescue (*Festuca thurberi*) were grazed more than 30% by weight, then utilization by weight removed was highly correlated to percent of plants grazed. Although the method is site specific, it can be used for local applications once the appropriate equation has been developed between percent grazed or ungrazed plants and percent utilization.

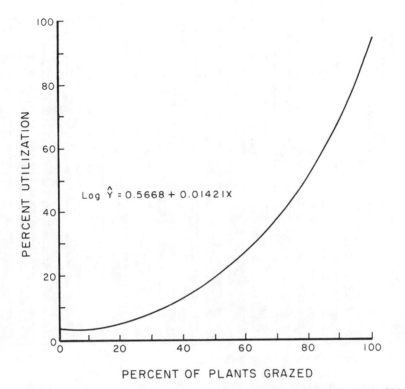

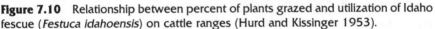

$$\text{Log } \hat{Y} = 0.5668 + 0.01421X$$

PERCENT OF PLANTS GRAZED

Figure 7.10 Relationship between percent of plants grazed and utilization of Idaho fescue (*Festuca idahoensis*) on cattle ranges (Hurd and Kissinger 1953).

7.8.7 Shortcut Method

The shortcut method is based on the occurrence of a high correlation among percentage of grasses grazed to a stubble height of 5 cm or less, the amount grazed above this height, and the ungrazed complement (Canfield 1942). It is called a "shortcut method" because plants grazed to a 5-cm stubble height can be easily noticed and the survey carried out quickly by one person who need not be a highly trained technician. It may be advisable, in some cases, to substitute the most frequently occurring grazed height for 5-cm stubble height. Percentage of tufts grazed to 5 cm or below is found from measurements using sample plots or line transects. For example, in a survey, 47% of the grass tufts were found to have been grazed. The number closest to 47 in the portion of the chart (Fig. 7.11) showing percent grazed at 5 cm or less is 50. Therefore, about 26% of the plants would be partly grazed and 24% ungrazed, in terms of basal tuft area. The method may not be applicable to tall, course-stemmed grasses where per-

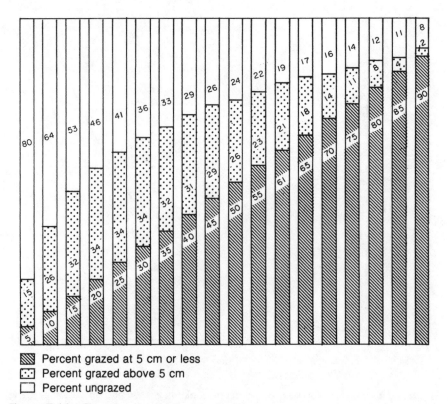

☒ Percent grazed at 5 cm or less
⋮⋮ Percent grazed above 5 cm
☐ Percent ungrazed

Figure 7.11 Chart for shortcut method of estimating grazing use [redrawn from Canfield (1942)].

manent stubbles are formed by a long history of grazing and the animals utilize only part of the current growth from the stubbles.

7.8.8 Stubble-Height and Line-Intercept Methods

The stubble-height and line-intercept methods evolved from the height–weight method and was developed by Canfield (1942, 1944) as part of a line-intercept procedure. Stubble-height classes are first determined for predominant grasses before the survey is conducted. Canfield (1942) divided stubble heights of grasses of similar growth form in southern Arizona into the following classes:

Class	Stubble Height (cm)
1	≤1.2
2	1.2–2.5
3	2.5–5.0
4	5.0–10.0

5	10.0–15.0
6	15.0–20.0
7	20.0–25.0
8	≥25.0
9	Ungrazed (any height)

For vegetation made up of tall, course-stemmed grasses, either the number of classes may be increased or larger class intervals may be used (Brown 1954). The height of each tuft, intercepted by a line transect, is measured from ground level, and the tuft is assigned to one of the predetermined classes. Height of grazed and ungrazed tufts is recorded to the nearest 2.5 cm and the horizontal extent of the tuft is measured at ground level. The intercept is recorded under the appropriate stubble-height class. If a tuft has been grazed to two different heights, the stubble height and the intercept of each grazed part of the plant is measured and recorded under the corresponding stubble-height class. The intercept by species across all stubble-height classes is summed, and percent composition of each species is determined as follows:

Percent composition of species

$$= \frac{\text{total intercept of species A}}{\text{total intercept of all species}} \times 100 \quad (7.5)$$

Then, percent intercept by stubble-height class is determined for each species:

Percent intercept by stubble-height class

$$= \frac{\text{total intercept in a stubble-height class of species A}}{\text{total intercept in all classes of species A}} \times 100$$

$$(7.6)$$

Results of a survey with several transects averaged for species will then give utilization estimates by species. Mean use is the weighted average for all species. This latter value is calculated by multiplying the percent intercept of each species in a stubble-height class by percent composition of that species. Total mean use in that class is the sum of the mean use for each species in that class.

A mean use of 14.7 for a stubble-height class, for example, is interpreted as 15% of stubble of all species had been grazed down to that stubble-height class. However, in order to interpret these results in terms of degree of utilization (heavy, moderate, or light), standards for comparison to a level of percentage removed to attain heavy use, for example, should be established. This method appears difficult at first sight, but once the standards are fixed it gives a complete view of utilization of vegetation. This method is applicable only to grass species and, generally, only perennial grasses are measured to assess utilization. The cost of survey by this method compared favorably with other methods (Canfield 1944), and, probably, this is still true.

7.9   INDIRECT BROWSE UTILIZATION METHODS

Estimation of browse utilization is one of the most difficult vegetation characteristics to determine, and some of the previous methods may not be efficient. Because of costs and destruction of plants, entire shrub or tree plants cannot be harvested to find differences in weights before and after grazing. Therefore, reliance has to be placed on indirect methods. Indirect methods are based on measurements of twig lengths and diameters, or counting numbers of grazed and ungrazed twigs.

7.9.1   Twig Length Method

Nelson (1930) was among the first to use twig lengths to estimate browse utilization. Current year's growth of twigs is measured on marked twigs in browsed and unbrowsed plots. The average difference in length of browsed and unbrowsed twigs is divided by average length of unbrowsed twigs to compute percent utilization. The method is analogous to that of height of grasses. One of the problems is that animals may eat only leaves without affecting twig length very much, if at all. Yet, leaves may constitute only 50–70% of twig weight. Therefore, utilization by this method would be underestimated. Another problem is that the relationship between twig length and weight is not always linear (Fig. 7.12). To alleviate this problem, use a quadratic regression instead of the linear form. That is, let

$$y = a + bx^2 \tag{7.7}$$

instead of $y = a + bx$, where y is the weight of twigs and x is the length of twigs. Or if percentage removal of weight is needed, then allow y to be percentage of weight removed when x is the percentage of twig length that has been removed. One would expect that biomass of a given twig is best expressed by relations presented in Chapter 6 on shrub and tree biomass. Recall the volume–weight curves of grasses used to estimate herbage removal. For instance, the volume of a twig or stem is a better estimator of weight than a single linear measure of length (height in volume measure). Then, first develop an equation of the general form

$$\text{Twig weight} = f \text{ (twig volume)}$$
$$= b(\pi r^2 \times \text{height of stem})$$

where b is the coefficient obtained from the regression method of least

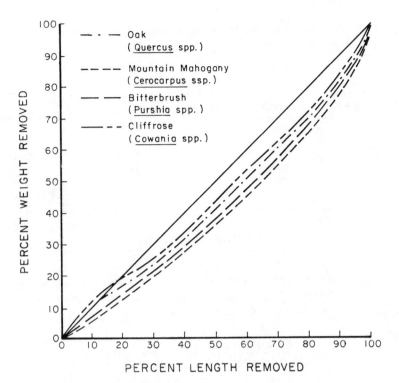

Figure 7.12 Relationship between length of leader removed and weight removed (Smith and Urness 1962).

squares. Since volume is usually of a conic shape (stem is tapered), then more specifically, the weight–volume relationship is

$$\text{Twig weight} = b(\tfrac{4}{3} \pi r^2) \tag{7.8}$$

where r is the stem radius at the base for the current year's growth.

7.9.2 Production Index Method

Length of the current year's growth is measured on 25–30 randomly selected twigs of a plant. On the same plant, the remaining length of current year's growth on the same number of browsed twigs is measured. The crown area of the plants is also measured. Average length of current year's twig growth is multiplied by the crown area (Brown 1954). This product is called the "production index" and is an estimate of total forage produced by a shrub or tree. The production index of each plant in a sampling unit are summed to obtain the total production index of that species. Similarly, an estimate of percentage utilization of twigs is calculated for each plant. A utilization index for each plant is obtained by multiplying the production

TABLE 7.10 Example of Production Index Computation to Determine Utilization on Woody Plants

1	2	3	4	5	6	7
	Crown Diameter		Ungrazed Twig	Production	Estimated Amount	Amount
Plant	(cm)	Area (cm²)	Length (cm)	Index	Removed	Grazed
1	38	1,134	50	56,700	20	11,340
2	76	4,534	38	172,292	20	34,458
3	50	1,962	30	58,860	25	14,715
4	30	706	64	45,184	15	6,778
5	38	1,134	40	45,360	70	31,752
Total[a]				378,396		99,043

From Hormay (1943b).

[a]Total percent utilization = $(99,043/378,396)100 = 26.2\%$.

index with percent utilization for that plant. A sum of utilization indices of all woody plants in the sample plot gives the "total utilization index." Percent utilization is then estimated as follows:

$$\frac{\text{Total utilization index}}{\text{Total production index}} \times 100 \tag{7.9}$$

The average percentage utilization of all sample plots then gives percent browse utilization of the area. Sample calculations for a sample plot are shown in Table 7.10.

The production index method was developed by Hormay (1943a,b) to estimate utilization and is suitable for only those woody species on which current growth is easily recognized. The procedure does not estimate the actual weight removed but, rather, is a relative index of weight. Aldous (1944) used only the number of browsed and unbrowsed twigs to develop an index of utilization. The number of browsed twigs divided by the total number of twigs counted gives the proportion of twigs utilized, which is also a relative index. No weight is involved, and these methods cannot be used to estimate carrying capacity. However, this method is useful for monitoring and evaluating utilization of all woody plants.

7.9.3 Branch Diameter

Twigs usually taper to a tip. Therefore, the diameter of a twig progressively decreases from its base to its tip. Then, it is possible to determine the relationship between the diameter at any one point and the amount of forage between the point and tip of the twig. Once the equation is determined, the amount of forage removed can be estimated by measuring the diameter where a twig has been cut off by grazing. Regression analysis in studies shows that a high correlation exists between diameter and weight of twigs in many species. Basile and Hutchings (1966) used two variables (twig diameter and length), simultaneously, to predict weight. One may always improve on prediction equations by adding information offered by other variables. However, the primary form of the relationship should always be determined; that is, determine whether the form is linear, quadratic, or exponential for each independent variable and the weight of twigs. Then, the equation will be more effective in predicting the weight remaining. One limitation of these methods is the wide variations in the diameter and diameter–length relationship to the weight of twigs from plant to plant and from year to year.

7.10 INSECT HERBIVORY

By definition, herbivores eat living plant tissues. This definition of herbivores includes not only large herbivores such as livestock and wildlife but also insects that feed on plant parts. The techniques discussed so far in this chapter for estimating herbage removal may also be used for other herbivores. The impact of less visible herbivores, such as grasshoppers and crickets, on native vegetation normally goes unnoticed unless insects assume an epidemic level. During severe outbreaks of insect populations, native vegetation may be heavily grazed. Most insect grazing goes beyond actual forage consumption. That is, insects often cut plant stems and blades yet eat only a part of the material cut.

For example, grasshoppers clip approximately six times as much foliage as they consume. One grasshopper (from hatching to 30 days of adult life) per square meter clips 14–17 kg/ha of forage (Hewitt 1977, 1978; Mulkern et al. 1969; Putnam 1962). One Canadian study on a needlegrass–wheatgrass prairie estimated that 16–60% of the total forage production was reduced by 10 grasshoppers per square meter during their life cycle (Hardeman and Smoliak 1982). Economic considerations for large herbivore production often require a quantitative assessment be made of herbage losses due to insect populations. However, insects foraging on plant roots or those sucking plant sap cause an indirect loss of plant productivity that cannot be quantitatively assessed easily by vegetation measurement techniques discussed in this book. Therefore, the techniques described here are for assessment of direct herbage removal by insects such as grasshoppers and crickets. Many examples exist for studies of foliage removed by grasshoppers. However, methods described may be used to measure foliage removed by most other insects (Bonham 1987).

7.10.1 Herbage Losses Due to Insects

It is extremely difficult to make an accurate estimate of the forage loss caused by insects. The major problem is the choice of a suitable method. For instance, insect foraging may result in a competitive advantage developing between grazed and ungrazed plants for space and nutrients. The reduced competition may increase the productivity of ungrazed plants to partially compensate for the losses. However, most methods to assess insect damage do not take this fact into consideration (Walker 1983).

Great variation in the biomass of plants of native vegetation due to microenvironmental differences is also a major source of error. However, this problem can be overcome to a certain extent by careful work and stratified sampling. Although most methods were developed for agronomic crops, methods discussed here have either been applied to native vegetation or have potential application.

7.10.2 Defoliation Classification Technique

The whole range of defoliation (0–100%) is classified into a number of categories. The extent of defoliation on individual plants is rated according to categories, and the number of plants in each category in a sample plot. Hewitt and Berdhl (1984) evaluated grasshopper preferences by this method. The following defoliation categories were used:

Rating	Category (Percent Leaves Destroyed)
0	0
1	1–25
2	26–50
3	51–75
4	76–99
5	100

The defoliation ratings of each plot are first weighted according to the number of plants in each category and then averaged to provide a single score. The number of defoliation categories may be increased for greater accuracy in assessment of damage. The procedure does not make a clear distinction between partially and fully damaged leaves. For more intensive work, categories may be weighted according to the extent of damage to individual leaves of plants.

7.10.3 Plot and Cage Methods

The plot and cage methods involve locating sample plots in areas with and without insect grazing. Plants within the sample plots are then clipped.

The difference in average oven-dry weights of the two sets of plots, from the area with and without insects, gives an estimate of forage removal by insects. This estimate may be obtained for total biomass or for individual species. In the latter case, the clipped plants of different species in each sample plot are kept separately. Sample plots should be located in pairs to reduce sampling error, and the number of pairs for statistical adequacy should be calculated from preliminary sampling. Follow Section 3.3, on methods, to estimate sample size for comparison of two means. In the present case, plots not grazed (caged) by insects provide an estimate of forage yield, while uncaged plots provide the same estimate minus forage removed by insects. Equation (3.44) provides an estimate for samples needed to detect a given level of differences in plant biomass between grazed and ungrazed areas.

Putnam (1962) used 0.4-m^2 plots to estimate the forage removal potential of grasshoppers in native grasslands of British Columbia. Plots were enclosed by 20-mesh screen cages, 0.7 m high. Insects in different densities are released into the cages. Foliage remaining on plots at the conclusion of the experiment is clipped close to the soil surface. Loss due to grasshopper feeding is estimated by subtracting the average oven-dry foliage weight of the grazed treatments from that of the control. The average consumption by a grasshopper is obtained by dividing the foliage yield removed by the average number of grasshoppers in each treatment.

Hewitt (1980) used cages to study the tolerance of 10 grass species to feeding by *Labops hesperius*. Nylon mesh is used to make two sizes of cages. The large-sized cages (305 × 152 × 122 cm) are used to enclose a sample plot containing a mixture of plants, while small-sized cylindrical cages (20-cm diameter × 122-cm height) are used to cover individual plants. Adult insects are introduced in both small and large cages to study plant tolerance to and food preferences of grasshoppers. Data collected to compare treatment effects include number of culms per plant, percentage of abnormal seed heads, seed production, seed weight, seed germination, forage production, and percentage leaf removed.

Anderson and Wright (1952) used cages with screen tops and sides. The cages measure 0.6 m long, 0.6 m wide, and 1.2 m high and are placed at random in the area to be measured approximately 2 hr after sunset or at such time when little or no grasshopper activity is detected. The bottom edge of each cage is banked with dirt to prevent the movement of grasshoppers in or out of the cages. The following morning, grasshoppers, which have been trapped by placement of the cages the previous evening, are removed from the cage through a door in one side, and the number and species are recorded. Cages are then removed from the area during the day to avoid any effect their presence might have on local grasshopper distribution.

All foliage in the plots is clipped to ground level, and three successive clippings are made at monthly intervals. Clipped foliage is air-dried and weighed and expressed in units desired, such as kilograms per hectare air-dry weight. By comparing the differences in weights of foliage on the three clipping dates, a measure of foliage removed is obtained. Monitoring of biomass loss follows the same procedures for time intervals as those given in caged plot methods in Section 7.11.

7.10.4 Feeding Trials

Estimates of vegetation consumed are sometimes arrived at indirectly by feeding plant materials to insects in laboratory cages. Mitchell (1975) used feeding trials to study variation in food preferences of three grasshopper species. Cages (56 × 30 × 24 cm) used are made of 1.27-cm plywood and window screening on three sides and tops. Shoots of different species are placed upright in vials in the cages. Grasshoppers are allowed to feed for 24 hr, and after removing the grasshoppers from cages, the amount of forage consumed is estimated on an oven-dry weight basis by subtracting the residual forage from initial available forage. Oma and Hewitt (1984) used a similar technique to estimate food consumption by grasshoppers.

7.10.5 Forest Canopy Defoliation

Clark (1961) mounted a pinhole camera with a wide-angle lens on a tripod and pointed it skyward to obtain 1) an index of interception of sky radiation, 2) index of back radiation of trees, 3) shading cover, 4) index of precipitation, and 5) shelter or wind exposure cover. Photos are taken at fixed points on a grid system in the forest. Tree heights and distances are determined using trigonometry, as given in Chapter 2. Forest canopy cover is determined from negatives with transparent grid overlays, and cover is obtained by 1) a dot counting system, 2) visual fractional estimate of cover in each segment, and 3) by combining light and dark areas of adjacent blocks of each segment. This technique is useful for studying the relationship between crown cover and survival of understory species, pre- and post-logging activity as it relates to canopy openings, and permanent records of shelter versus exposure of weather stations. Repeated use of the technique over time could be used to evaluate amount of defoliation occurring because of disease, insects, and so forth.

Studies have been conducted of gypsy moth populations in forests and their control by airplane spraying. Plots from 10–80 ha/ac in size (averaging 28 ha) were located on topographic maps; subplots of 0.04 ha (20 × 20 m) were used as sampling units. Prior to egg hatch and treatment, visual estimates of egg mass were made by walking through subplots. A second method employed counting egg masses on individual trees, on other plants, and on all ground litter in subplots. The same methods were used after spray treatments. Live larval counts were made as observers walked along a line between sub-plots. Defoliation readings were made using binoculars and estimating percentage of leaf defoliated. Data was ranked using Spearman's Rank Correlation Coefficient.

Cooper et al. (1987) used circular plots (0.04 ha) and a sighting tube (4 cm in diameter) with a cross-hair sight to obtain estimates of defoliation in tree canopies. Canopies were divided into 0–6, 6–12, 12–18, 18–24, and >24-m height classes. Hits or misses of foliage on the cross-hair in each height class were recorded. Five point-intercept samples were taken in each compass direction, and the number of hits in each height class divided by the number of sample points ($n = 20$) is an estimate of the percentage canopy cover for each height class. Samples were taken once before defoliation and in the same location at the peak defoliation period to estimate amount of defoliation. The point-intercept method of defoliation estimation is objective rather than subjective. However, the method can be modified to estimate defoliation on individual trees. In this case, the tree stem is treated as the plot center and sampling points radiate outward from it. Variation exists among observers as to where cross-hairs are placed when viewing upper foliage after viewing lower foliage which may block the view of upper foliage. Variation also exists in placing of sample points from one sighting location to another. Difficulty arises when one tries to determine partial defoliation on an individual leaf in the upper canopy.

7.11   CAGED PLOTS FOR UTILIZATION

Most measurements of forage removal employ the use of cages. Both herbaceous and browse species can be caged. Cages or other types of exclosure are used extensively to protect vegetation from herbivores to determine forage removal or to study vegetation composition, plant growth, height, and other characteristics. Paired plots, one enclosed and the other in an area open to herbivory, are used to measure the amount of forage removed by grazers.

Is there a significant effect of cage or exclosure on the enclosed vegetation? The effects of cages on production of biomass may be different at various times of the day and with various combinations of weather. Growth is often greater within the cage than outside, and the presence of cages may cause a small increase in plant growth early in the growing season. However, these differences soon disappear as temperatures become warm enough for faster plant growth. According to several studies, no significant differences have been found in species composition or decomposition rates inside versus outside the cages. According to Owensby (1969), during the first few years of an experiment using cages, areas grazed heavily or lightly by livestock would tend to have a greater relative difference between caged and uncaged areas than on moderately grazed pastures. Although the cage effect limits the value of cages to measure utilization of foliage, alternative methods also have limitations. A specific knowledge of cage effects on herbage growth over a period of years may be needed to make adequate corrections to determine amounts of true grazing use only if herbivory is small compared to effects of cages on biomass production.

Productivity (the amount of biomass produced over a given time interval) also employs the use of cages to prevent grazing of plant materials. If grazing is not present, which is usually not the case, then, of course, cages are not needed. Then, one only needs to employ the relevant measurements over the time intervals desired and use the difference methods presented here for the use of cages.

Let

$$U = P_c - P \tag{7.10}$$

where U is the amount of forage used, P_c represents the caged plot biomass weight, and P is the uncaged plot biomass weight. To gain maximum efficiency in sampling, pair the caged and uncaged plots over all observations. First, select a random point to place a cage. Then, select a plot that is as close as possible to that caged in species composition, cover, and biomass. The second plot should also be as close in proximity to the caged plot as possible. Thus, variations in vegetation, soils, topography, and microclimate are minimized.

At the end of the selected time interval, both pairs are harvested and averaged over all observations to estimate U. If more than one time interval is needed or desired, then

$$U = \sum_{i=1}^{T} (P_{ci} - P_i) \tag{7.11}$$

where P_{ci} represents the average amount of biomass in caged plots at time i and P_i is the amount of biomass in uncaged plots at time i. Note that measurement of utilization for biomass at $t = 2$, P_i becomes a caged plot, and the new difference is regrowth since the grazed plot in $t = 1$ is now caged. Large sample sizes ($n > 25$) should be used to ensure that no negative differences occur. That is, at least 25 pairs of plots of sufficient size should be used over each time interval. If soil, topographic, or other environmental features vary, then the area should first be stratified and sampled according to strata. Each strata may need variable sample sizes for adequacy.

The variance of U in Equation (7.11) is

$$V(U) = \sum_{i=1}^{t} [V(P_{ci} - P_i)] + 2 \text{ Cov } [(P_{ci} - P_i)(P_{cj} - P_j)] \tag{7.12}$$

for $i \neq j$.

Total amount of biomass B produced is

$$B = P_i + \sum_{i=1}^{t} (P_{ci} - P_i) \tag{7.13}$$

where P_i is the initial amount of biomass at the beginning when cages are placed. It is obvious that $P_i = P_{ci}$ for the first time interval. Then, rewriting Equation (7.13), we obtain

$$B = P_{c1} + \sum (P_{c2} - P_{it1})$$

Percentage utilization for a season is $(U/B) \times 100$.

7.12   MODELS VERSUS MEASUREMENTS FOR MONITORING

Measurement of vegetation for the purpose of monitoring requires a high level of accuracy and repeated measurements. The availability of financial and technical resources may become serious constraints for measurement of acceptable levels of statistical error. A good solution to the problem could be modeling, but the ability of models to adequately follow field data is limited. A combination of model results and field data, however, frequently yields an estimate with satisfactory statistical errors (Jameson 1986a). The combination may allow more precise interpretation of the results than is possible from measurements alone. For details of the procedure, interested readers with appropriate statistical background should refer to the work of Jameson.

Monitoring is a costly procedure, but it is possible to reduce cost by comparing the cost of a small sample taken frequently with the cost of a less frequent large sample. The combination of field data with model results can be used to predict the state of the system in the intervening period. Details of these procedures are beyond the scope of this book, and for details, interested readers should refer to Jameson (1985, 1986b).

7.13   BIBLIOGRAPHY

Aldous, S. E. 1944. A deer browse survey method. *J. Mammol.* **25:** 130–136.

Anderson, M. C. 1964. Studies of the woodland light climate. I. The photographic computation of light conditions. *J. Ecol.* **52:** 27–41.

Anderson, N. L., and J. C. Wright. 1952. *Grasshopper Investigations on Montana Rangelands.* Montana Agriculture Experiment Station Bulletin 486.

Austin, M. P. 1981. Permanent quadrats: An interface for theory and practice. *Vegetatio* **46:** 1–10.

Basile, J. V., and S. S. Hutchings. 1966. Twig diameter-length–weight relations of bitterbrush. *J. Range Manage.* **19:** 34–38.

Beeftink, W. G. 1979. Vegetation dynamics in retrospect and prospect: Introduction to the proceedings of second symposium of the working group on succession research on permanent plots. *Vegetatio* **40:** 101–105.

Bonham, C. D. 1987. Estimation of forage removal by rangeland pests. In *Integrated Pest Management on Rangeland—A Shortgrass Prairie Perspective*, J. L. Caperina (ed.). Westview Press, Boulder, CO, 426 pp.

Brown, D. 1954. *Methods of Surveying and Measuring Vegetation*. Bulletin 42. Commonwealth Bureau of Pastures and Field Crops, Hurley, Berkshire, 223 pp.

Buckman, R. E., and C. Van Sickle. 1983. *Resource change information is the key to Forest Service planning*, pp. 24–27. *In* Proceedings of International Conference on Renewable Resource Inventories for Monitoring Changes and Trends (August 15–19, 1983), J. F. Bell and T. Atterbury (eds). Oregon State University, Corvalis. 737 pp.

Caird, R. W. 1945. Influence of site and grazing intensity on yields of grass forage in the Texas Panhandle. *J. Forestry* **43:** 45–49.

Campbell, R. S. 1943. Progress in utilization standards for western ranges. *J. Wash. Acad. Sci.* **33:** 161–169.

Canfield, R. H. 1942. *A Short-cut Method for Checking Degree of Forage Utilization.* Research Note 99, USDA Forest Service, Southwestern Forest and Range Experiment Station, 5 pp. (mimeo.).

———. 1944. Measurement of grazing use by the line interception method. *J. Forestry* **42:** 192–194.

Cassady, J. T. 1941. A method of determining range forage utilization by sheep. *J. Forestry* **39:** 667–671.

Clark, F. G. 1961. A hemisperical forest photocanopy meter. *J. Forestry* **59:** 103–105.

Clark, I. 1945. Variability in growth characteristics of forage plants on summer range in central Utah. *J. Forestry* **43:** 273–283.

Connola, D. P., F. B. Lewis, and J. McDonough. 1966. Experimental field techniques used to evaluate gypsy moth (*Porthetria dispar*) control in New York. *J. Econ. Entomol.* **59:** 284–287.

Cook, C. W., and L. A. Stoddart. 1953. The quandary of utilization and preference. *J. Range Manage.* **6:** 329–335.

Cooper, R. J., K. M. Dodge, and R. C. Whitmore. 1987. Estimating defoliation using stratified point intercept sampling. *Forest Sci.* **33:** 157–163.

Crafts, E. C. 1938. Height–volume distribution in range grasses. *J. Forestry* **36:** 1182–1185.

Culley, M. J., R. S. Campbell, and R. H. Canfield. 1933. Values and limitations of clipped quadrats. *Ecology* **14:** 35–39.

Dyksterhuis, E. J. 1948. The vegetation of the Western Cross Timbers. *Ecol. Monogr.* **18:** 325–376.

———. 1949. Condition and management of rangeland based on quantitative ecology. *J. Range Manage.* **2:** 104–115.

Gardiner, H. G., and B. E. Norton. 1983. Do traditional methods provide a reliable measure of range trend, pp. 618–622. In *Proceedings of International Con-*

ference on Renewable Resource Inventories for Monitoring Changes and Trends (August 15–19, 1983), J. F. Bell and T. Atterbury (eds.). Oregon State University, Corvallis.

Gierisch, R. K. 1967. An adaptation of the grazed plant method for estimating utilization of thurber fescue. *J. Range Manage.* **20:** 108–111.

Goodwin, W. F., and J. Walker. 1972. *Photographic Recording of Vegetation in Regenerating Woodland*. Woodland Ecology Unit, Commonwealth Scientific and Industrial Research Organizations, Canberra City, Australia. Technical Communication No. 1.

Hardeman, J. M., and S. Smoliak. 1982. *Impact of Grasshoppers on Rangelands*. Research Highlights, Canada Research Station, Lethbrige, pp. 59–62.

Harper, J. L. 1977. *Population Biology of Plants*. Academic Press, London, 892 pp.

Hattersley-Smith, G. 1966. The symposium on glacier mapping. *Can. J. Earth Sci.* **3:** 737–743.

Heady, H. F. 1950. Studies on bluebunch wheatgrass in Montana and height–weight relationships of certain range grasses. *Ecol. Monogr.* **20:** 55–81.

———. H. F. 1975. *Rangeland Management*. McGraw-Hill, New York, 460 pp.

Hewitt, G. B. 1977. *Review of Forage Losses Caused by Rangeland Grasshoppers*. USDA Miscellaneous Publication No. 1348, 22 pp.

———. 1978. Reduction of western wheatgrass by the feeding of two rangeland grasshoppers *Aulocara elliotti* and *Melanoplus infantilis*. *J. Econ. Entomol.* **71:** 419–421.

———. 1980. Tolerance of ten species of Agropyron to feeding by *Labops hesperius*. *J. Econ. Entomol.* **73:** 779–782.

Hewitt, G. B., and J. D. Berdahl. 1984. Grasshopper food preference among alfalfa cultivars and experimental strains adapted for rangeland interseeding. *Environ. Entomol.* **13:** 828–831.

Hocker, R. B. 1978. Use of large scale aerial photographs for ecological studies in a grazed arid ecosystem, pp. 1895–1908. In: *Proceedings of 12th International Symposium on Remote Sensing of Environment*. Environmental Research Institute of Michigan and Natural Resources Management Centre, Phillipines.

Hormay, A. L. 1943a. *Bitterbrush in California*. Research Note 34, USDA Forest Service, California Forest and Range Experiment Station, 13 pp.

———. 1943b. *A Method of Estimating Grazing Use of Bitterbrush*. Research Note 35, USDA Forest Service, California Forest and Range Experiment Station, 4 pp.

Hurd, R. M., and N. A. Kissinger, Jr. 1953. *Estimating Utilization of Idaho Fescue (Festuca idahoensis) on Cattle Range by Percent of Plants Grazed*. Rocky Mountain Forest and Range Experiment Station, Paper No. 12, pp. 1–5.

Jameson, D. A. 1985. A priori analysis of allowable interval between measurements as a test of model validity. *Appl. Math. Comput.* **17:** 93–105.

———. 1986a. Models versus measurements in grazing systems analysis, pp. 37–41. In *Symposium Proceedings. Statistical Analysis and Modelling of Grazing Systems Data* (February 9–16, 1986), C. D. Bonham, S. S. Coleman, C. E. Lewis, and G. W. Tanner (eds). Society of Range Management 39th Annual Meeting, Orlando, FL.

———. 1986b. Sampling intensity for monitoring of environmental systems. *Appl. Math. Comput.* **18:** 71–76.

Kleiber, M. 1961. *The Fire of Life: An Introduction to Animal Energetics.* Wiley, New York, 454 pp.

Lacaze, B., G. Debussche, and J. Jardel. 1983. Monitoring changes and trends with LANDSAT and ancilliary data: Example taken from Mediterranean Lands, pp. 39–41. In *Proceedings of International Conference on Renewable Resource Inventories for Monitoring Changes and Trends* (August 15–19, 1983), J. F. Bell and T. Atterbury (eds.). Oregon State University, Corvallis.

Law, R. 1981. The dynamics of a colonizing population of *Poa annua. Ecology* **62:** 1267–1277.

Lommasson, T., and C. Jensen. 1938. Grass volume tables for determining range utilization. *Science* **87:** 444.

Mitchell, J. E. 1975. Variation in food preferences of three grasshopper species (Acrididae: Orthoptera) as a function of food availability. *Am. Midl. Nat.* **94:** 267–283.

Mosley, J. C., S. C. Bunting, and M. Hironaka. 1986. Determining range condition from frequency data in mountain meadows of Central Idaho. *J. Range Manage.* **39:** 561–565.

Mulkern, G. B., K. P. Pruess, H. Knutson, A. F. Hogen, J. B. Campbell, and J. D. Lambley. 1969. *Food Habits and Preferences of Grassland Grasshoppers of North Central Great Plains.* North Dakota State University Agriculture Experiment Station Bulletin No. 481, 32 pp.

Nelson, E. W. 1930. Methods of studying shrubby plants in relation to grazing. *Ecology* **11:** 764–769.

Oma, E. A., and G. B. Hewitt. 1984. Effect of *Nosema locustae* (Microsporida: Nosematidae) on food consumption in the differential grasshopper (Orthoptera: Acrididae). *J. Econ. Entomol.* **77:** 500–501.

Owens, M. K., H. G. Gardiner, and B. E. Norton. 1985. A photographic technique for repeated mapping of rangeland plant populations in permanent plots. *J. Range Manage.* **38:** 231–232.

Owensby, C. E. 1969. Effect of cages on herbage yield in true prairie vegetation. *J. Range Manage.* **22:** 131–132.

Parker, K. W. 1951. *A Method for Measuring Trend in Range Condition on National Forest Range.* USDA Forest Service, 26 pp.

Pechanec, J. F. 1936. Comments on the stem-count method of determining the percentage utilization or range. *Ecology* **17:** 329–331.

Pickford, G. D. 1940. Range survey methods in western United States. *Herbage Rev.* **8:** 1–12.

Putnam, L. G. 1962. The damage potential of some grasshoppers (Orthoptera: Acrididae) of the native grasslands of British Columbia. *Can. J. Plant Sci.* **42:** 596–601.

Ratliff, R. D., and S. E. Westfall. 1973. A simple stereographic technique for analyzing small plots. *J. Range Manage.* **26:** 147–148.

Reid, E. H., and G. D. Pickford. 1941. A comparison of the ocular-estimate-by-plot and the stubble-height methods of determining percentage utilization of range grasses. *J. Forestry* **39:** 935–941.

Roach, M. E. 1950. Estimating perennial grass utilization on semi-desert cattle range by percentage of ungrazed plants. *J. Range Manage.* **3:** 182–185.

Rogers, G. F., R. M. Turner, and H. E. Malde 1983. Using matched photographs to monitor resource change, pp. 90–92. In *Proceedings of International Conference on Renewable Resource Inventories for Monitoring Changes and Trends* (August 15–19, 1983), J. F. Bell and T. Atterbury (eds.). Oregon State University, Corvallis.

Sampson, A. W. 1919. *Plant Succession in Relation to Range Management.* USDA Bulletin No. 791.

Sickle, C. V. 1983. Measuring resource change in a changing world, pp. 36–38. In *Proceedings of International Conference on Renewable Resource Inventories for Monitoring Changes and Trends* (August 15–19, 1983), J. F. Bell and T. Atterbury (eds.). Oregon State University, Corvallis.

Smith, A. D., and P. U. Urness. 1962. *Analyses of the Twig Length Method of Determining Utilization of Browse.* Utah State Department of Fish and Game, Publication No. 62, 9 pp.

Springfield, H. W., and G. Peterson. 1964. *Use of the Grazed Plant Method for Estimating Utilization of Some Range Grasses in New Mexico.* USDA Forest Service, Rocky Mountain Forest and Range Experiment Station Research Note RM-22.

Stewart, G., and S. Hutchings, 1936. The point-observation plot (square-foot density) method of vegetation survey. *J. Am. Soc. Agron.* **28:** 714–722.

Stoddart, L. A. 1935. Range capacity determination. *Ecology* **16:** 531–533.

Stoddart, L. A., A. D. Smith, and T. W. Box. 1975. *Range Management.* McGraw-Hill, New York, 532 pp.

USDA Soil Conservation Service. 1967. *National Range Handbook.*

USDA. 1982. *Range Analysis Handbook.* Region IV, U.S. Forest Service, Ogden, Utah. Section 4.63, 46 pp.

Walker, P. T. 1983. Crop losses: The need to quantify the effects of pests, diseases and weeds on agricultural production. *Agriculture Ecosystems and Environment* **9:** 119–158.

West, N. E. 1983. Choice of vegetation variables to monitor range condition and trend, pp. 636–639. In *Proceedings of International Conference on Renewable Resource Inventories for Monitoring Changes and Trends* (August 15–19, 1983), J. F. Bell and T. Atterbury (eds.). Oregon State University, Corvallis.

Winward, A. H., and G. C. Martinez. 1983. Nested frequency—an approach to monitoring trend in rangeland and understory timber vegetation, pp. 632–635. In *Proceedings of International Conference on Renewable Resource Inventories for Monitoring Changes and Trends* (August 15–19, 1983), J. F. Bell and T. Atterbury (eds.). Oregon State University, Corvallis.

APPENDIX
UNIT CONVERSION TABLES

TABLE A.1 Conversion Constants from the English System to Metric Units

From			To	
Length				
in	inches	2.54	centimeters	cm
ft	feet	0.30	meters	m
mi	miles	1.6	kilometers	km
Area				
in^2	square inches	6.5	square centimeters	cm^2
ft^2	square feet	0.09	square meters	m^2
mi^2	square miles	2.6	square kilometers	km^2
ac	acres	0.4	hectares	ha
Mass (Weight)				
oz	ounces	28	grams	g
lb	pounds	0.45	kilograms	kg
Volume				
fl oz	fluid ounces	30	milliliters	ml
pt	pints	0.47	liters	l
qt	quarts	0.95	liters	l
gal	gallons	3.8	liters	l
ft^3	cubic feet	0.03	cubic meters	m^3

TABLE A.2   Constants for Converting Weight/ Small Area to Weight/Large Area

From g/area	Constant	To
0.1 m^2	10	g/m^2
0.5 m^2	2	g/m^2
0.1 m^2	100	kg/ha
0.25 m^2	40	kg/ha
0.5 m^2	20	kg/ha
1.0 m^2	10	kg/ha
0.1 m^2	89.2	lb/acre
0.25 m^2	35.7	lb/acre
0.50 m^2	17.8	lb/acre
1 m^2	8.9	lb/acre
0.96 ft^2	100	lb/acre
9.6 ft^2	10	lb/acre
96 ft^2	1	lb/acre

INDEX